T0265538

SAVE OUR SOULS

THE TRUE STORY OF A CASTAWAY FAMILY, TREACHERY, AND MURDER

MATTHEW PEARL

HARPER

An Imprint of HarperCollinsPublishers

SAVE OUR SOULS. Copyright © 2025 by Matthew Pearl. All rights reserved. Printed in the United States of America. No part of this book may be used or reproduced in any manner whatsoever without written permission except in the case of brief quotations embodied in critical articles and reviews. For information, address HarperCollins Publishers, 195 Broadway, New York, NY 10007.

HarperCollins books may be purchased for educational, business, or sales promotional use. For information, please email the Special Markets Department at SPsales@harpercollins.com.

FIRST EDITION

Designed by Michele Cameron
art © Shutterstock/254280373,1993167239

Library of Congress Cataloging-in-Publication Data has been applied for.

ISBN 978-0-06-333806-7

24 25 26 27 28 LBC 5 4 3 2 1

For Petey, and all the pets who rescue us daily

Tonight, at sea, overtaken by a violent tempest by the will of Providence, full of despair and distress, kneeling before Him who could save our lives, and may, perhaps, be willing to save our souls, having nothing more to hope from men, but everything to fear from God.

—Victor Hugo, *The Man Who Laughs*

CONTENTS

SAVE OUR SOULS

PROLOGUE

February 3, 1888

WITH TIME RUNNING OUT, ELIZABETH Walker desperately searched for her family dog. She was trying to stay upright while crossing the swaying sailboat, as her husband and three teenage sons prepared for the worst. The *Wandering Minstrel* was a schooner, a sailing vessel with multiple masts, that had been anchored at a coral reef in the North Pacific Ocean. Once typhoon-force gales began, they could no longer see beyond the taffrail—nothing except the terrifying waves, illuminated by flashes of lightning.

The chains of the anchor snapped and the schooner now tossed about, and then dashed onto the jagged reef. The *Minstrel* had once represented an upturn in the Walker family's fortunes as well as a temporary home, or as poet Epes Sargent described sailing in childhood: "A life on the ocean wave, / A home on the rolling deep." Now they watched that home fall to pieces, and their only chance to survive was to abandon it. They hurried, along with their crew, to prepare the lifeboats, which is when Elizabeth realized there was still no sign of their beloved dog, who could have been anywhere—probably hiding.

"Jessie!" Elizabeth called, a call repeated by the other Walkers. "Jessie!"

Their voices were swallowed by the violent sounds of the sea and storm, and the family had to shout to each other to be heard as the swells crashed into the hull. Elizabeth needed warmer clothes to prevent hypothermia later once she was exposed to the elements. She pulled on a pair of her husband Frederick's trousers, a symbolic affront to those who believed a woman never belonged in a man's world of sailing in the first place. She tied a felt hat onto her head with a handkerchief so that it would not be blown off. Then she gathered up half a dozen heavy jackets that her family could use as layers.

Freddy, Henry, and Charlie Walker, who found no trace of Jessie, helped shipmates as directed by their father—the ship's captain, Frederick Walker. The boards of the schooner cracked and split. If they waited too long, they would be thrown right into the ocean and swept away.

Three lifeboats were lowered into the water. Charlie's eyes fell on "a succession of raging breakers." Giant, unmistakable shadows circled below the waves, with triangular fins breaking the surface. The sharks were anything but unexpected. In fact, Frederick had set course for this remote atoll—a geological formation of coral— partly to fish for sharks. Now, as the Walkers and crew faced the billows of the sea, the hunters were hunted.

Among the sharks that circled these reefs were those later classi- fied as Galapagos sharks, distinguished by the ridge between their dorsal (back) fins. They measured up to ten feet long, flashing fourteen rows of serrated teeth as they launched at prey. Though not inherently interested in humans, they lowered their fins into a hunched position and entered a kind of frenzy when threatened. More dangerous were the tiger sharks, solitary hunters weaving through the reef. Named for the stripes running along their flanks, larger tiger sharks were twenty feet long and weighed two thousand pounds, graceful yet voracious as they ate everything in sight, each with four dozen notched, curved teeth that could slice through prey.

The Walkers now had to rappel down lines into the lifeboats, resigned to leaving behind treasured possessions, as well as poor, lost Jessie. Elizabeth gripped the line and braced herself. Along with her family and shipmates, she was entering a new phase. The Walkers and their crew stood at the cusp of a life-altering ordeal that would come to be seen as a true-life *Robinson Crusoe*, with more than a touch of a real *Swiss Family Robinson*. With good reason, a Honolulu newspaper would come to call the events "a shipwreck and existence on a desert island as thrilling as is to be found in the annals of the sea."

Elizabeth's role was singular, not because she was the one woman there, but because she was seen as a kind of mother to all—the person looked to for moral and spiritual strength, which was, no doubt, an unfair burden that would test her. Vital questions accompanied them off the schooner: How would each of them transform in crisis, and how far would the Walkers have to go to protect their family?

As Elizabeth descended, a crew member gave the signal for her to let go of the line and drop into the boat. But the signal came a second or two prematurely, before the small craft was stable— which almost sent her into the shark-laden water.

She had just barely managed to touch down into the lifeboat before she was joined by the rest of her family. Then she heard a faint sound from the ship that made her heart leap: barking.

Chapter 1

THE VOYAGE BEGINS

Fall 1887

SIX MONTHS EARLIER, ELIZABETH WALKER, forty-three, was living in Hong Kong, with no reason to suspect her life would be forever changed as she prepared to go to sea again with Frederick. Her frequent voyages—often to locales many people never heard of, much less imagined visiting—were prefigured by her childhood. Liverpool, where she was raised and where her father taught music, was a bustling shipping and trading port, among the busiest in Europe. "The forests of the globe are represented, as in a grand parliament of masts," Herman Melville wrote of Liverpool's harbor in his novel *Redburn*, drawing on his own time there. With Elizabeth's parents dead by the time she was a young adult, she and two of her brothers, J. L. and Richard, embraced degrees of wanderlust, taking voyages around the world.

Elizabeth had married Frederick Walker, forty-eight, when they were in their early twenties. Originally from Dublin, Frederick was tall and slender, with a bushy mustache and a dry sense of humor. He relished spending time with Elizabeth, with whom he enjoyed a tender camaraderie, but as an officer on ships he was stern and no-nonsense. When he reflected that "it was my invariable custom to get up at night, go on deck, perhaps light a pipe, and feel the invigorating influence of the cool night air," he was describing himself not as a wizened old mariner, as his language conjures, but as a man in his thirties. His long spindly legs could appear awkward, even clumsy, as he climbed between decks. To casual observers he could seem ill-tempered and stringent, and he was the first to admit he demanded "apple pie order." One former apprentice praised him as good a sailor "as ever trod a deck." Frederick also looked for enterprising ways to build upon his expertise, including applying for a British patent for a mechanism for hydraulic power. Between his voyages as captain and his business ideas, opportunities had led the couple to live in Yokohama, Japan, and now Hong Kong.

Hong Kong, a British colony for forty-five years after becoming a focal point for disputes between China and Great Britain over the sale of opium, was a cosmopolitan mix of residents and visitors in pursuit of international commerce, not unlike the Liverpool of Elizabeth's childhood. Making preparations for a passage while in Hong Kong would take the Walkers through what one travel writer described as the "granite quays fringing the lake-like harbor, the tall merchant offices and cavernous ware[houses] beyond, the solidly-constructed buildings . . . climbing tier over tier far up the steep hillside."

Elizabeth was not the only woman of the era making long ocean passages with her family. A whaling captain's wife earlier in the nineteenth century recalled an emotional decision made when her husband sailed shortly after the birth of their child: "The separation seemed worse than before, and I then came to the determination that if he ever went to sea again I would accompany him." Still, women on commercial and fishing expeditions remained rare, and rarer was a woman who embraced the opportunities the way Elizabeth did.

Some women, after joining a grueling voyage, avoided repeating the experience. Elizabeth's fervor, by contrast, mirrored that held by the whaler's wife on her second morning at sea: "I came on deck, and all was one boundless expanse of ocean; the impression was one of magnificence."

Elizabeth's role on ships required an adventurous spirit and a willingness to challenge stereotypes. The shipping press noted with surprise that wives on board fishing vessels appeared to increase productivity in the crew, though observers seemed conflicted about the reason, surmising that women had "good influence" on discipline and morale. Being the one woman aboard was never easy, though, requiring composure and leadership amid the rowdy, overworked, and sometimes disgruntled men who made up the crews. Though the era had discarded many earlier superstitions about women at sea, if sailors thought a woman made too many demands, they were quick to label her cruel and overbearing. These were some of the usual obstacles as Elizabeth readied to set off on her latest passage with her husband in the fall of 1887.

By the time the Walkers had settled in Hong Kong, Frederick had lost most of their capital in unprofitable ventures. Still, a consortium of Hong Kong–based investors were impressed with Frederick's reputation for integrity after receiving strong endorsements from the Yokohama business community. That led them to commission him to captain a sharking vessel. As the Walkers finalized their plans to leave Hong Kong, some captains might have had second thoughts about the schooner secured for him, the *Wandering Minstrel*. This green, three-masted vessel seemed to attract bad luck, or, as the local newspaper in the town where it was built put it, was "dogged by ill fate from the day she was launched." Three years earlier, sailing out of Japan, the schooner lost its topmast in a typhoon and was declared a total loss. A number of the vessel's owners who were not properly insured were reported to be financially ruined. Frederick, though, mused that while some "sailors believe in lucky and unlucky ships," he did not. Besides, he could not afford superstitions.

Whatever others might have thought of the ship, Frederick was confident enough in his sailing skills not only to proceed but also to once again bring along Elizabeth and their three sons. For Elizabeth and Frederick, family meant a journey together rather than a gradual process of watching offspring disperse into separate corners of an industrializing society. But that togetherness came with unique pressures. The more time the children spent on vessels, the less education they received elsewhere, which meant greater onus on their parents to teach them what they needed in life. These duties fell mostly on Elizabeth. They were also personally responsible for expenses related to the children, separate from the voyages' business-related costs, which made profit harder to secure. Still, the *Minstrel* represented their best prospect to reverse their financial outlook, and in doing so to ensure future voyages.

The Walker boys knew no other life. They grew up as citizens of the world. By the time they looked forward to boarding the *Wandering Minstrel*, Frederick (Freddy) was seventeen, Henry, fifteen, and Charlie, fourteen, all of them slender and tall like their father. Blue-eyed Charlie recalled his first time at sea as being "when I was five months old," with much of childhood spent on ships, seeing different countries, hearing new languages, and trying unfamiliar foods. On Pacific islands, they could choose delicacies from "oranges, bananas, pineapples, plantains, breadfruits, coconuts, lemons, mummy apples, limes, guavas." When very young, the Walker boys became accustomed to being "tossed from one side to the other without the least mercy" while confronting natural fears about being out on the ocean. Another child who sailed in the era would peer out the cabin window, asking her doll, "Shouldn't you think I would be afraid of those deep, deep billows? Well, I *ain't*, for I know there's somebody up in heaven that will take care of me." Gestures toward normal routines and benchmarks also helped quell nerves. During Christmas, the young passengers enjoyed stockings hung for them, and on their birthdays crew members gave them small trinkets or coins.

Before they embarked on this latest journey, the Walker sons climbed around the *Minstrel*, the latest vessel that would serve as a combination schoolhouse, apprenticeship, and playground. Also aboard was the family's inquisitive dog, Jessie, whom the Walkers would not consider leaving behind. The boys gravitated toward the weapons on board, especially a particular cannon. Their father had spotted the relic lying in the dockyard at Hong Kong. In the hollow of the cannon the boys had found the letters GS engraved—was it short for Georgius Secundus, or King George II, and might the cannon have belonged to Great Britain's eighteenth-century navy? The dockmaster said it had been recovered with other pillaged items from a Chinese pirate junk (or sailing vessel), and had been on a British naval vessel that had taken it from a Spanish frigate, and before that might have come from any corner of the ocean. Charlie judged it a "very ancient design." They did not have the funds to pay for a nonessential item. But given the boys' interest, the dockmaster had offered it as a gift, as long as they could find a way to transport the four-hundred-pound object. Charlie rigged a carriage to pull it on.

Frederick Walker was one of a relatively small coterie of captains trying to establish a shark-fishing industry (also known as a fishery) around the Hawaiian islands, an area that did not have a long-standing history of organized fishing for sharks. Robert Louis Stevenson and his stepson Lloyd Osbourne's novel *The Wrecker* described these vessels in the burgeoning industry as conducting a "holy war on sharks." Sharking came with potential high rewards but ample uncertainty. In terms of demand, the unpredictable commercial value of products derived from sharks was set in Asia, and in terms of supply, sharks that seemed abundant at times across the Pacific would become scarce at other junctures, complicating the economics.

The *Wandering Minstrel*'s name evoked performers in medieval times who traveled from place to place to entertain and sing, but specifically paid homage to a popular one-act Victorian play of the same name by Charles Dickens's friend Henry Mayhew, about a man pretending to be a performer in order to take advantage of a wealthy family—a variation on the fact that Mayhew himself hid his playwriting from his respectable family. (The word *minstrel* separately developed racial connotations in the United States, which began to influence uses of the word abroad.) The *Minstrel* departed from Hong Kong with a modest-size crew on Thursday, October 13, 1887. Frederick would later remember this incorrectly as falling on Friday the thirteenth, an unsurprising error considering how events unfolded. Errant memory aside, many sailors thought any Friday, believed to be the day of Christ's Crucifixion, unlucky. The ship almost immediately faced gales and a typhoon that they sailed through approximately three hundred miles west of Formosa (now Taiwan). The storms damaged the steering mechanism and ruined provisions.

After the typhoon, Frederick sailed to Honolulu to arrange repairs while hoping that the weather would improve. At Honolulu, a reporter from the island's *Daily Bulletin* opined that the ship had "an appropriate name" after the Walkers invited him aboard while at anchor to hear music and singing. It was natural that Elizabeth, the daughter of a music teacher, encouraged music on

board the ship, and the festivities could include the Walkers play-
ing on the piano they brought along, as well as singing by crew
members.

At that time, the plans of all merchant ships including the
Minstrel were subject to the vicissitudes of the Kingdom of Hawaii's
politics. The unified nation of islands had been independent for
nearly a century. However, the sway of American plantation owners
in Hawaii had dramatically increased after a series of treaties elim-
inated tariffs for sugar exports to the United States, which in turn
required an increase in non-Hawaiian workers that engulfed the
native labor force. As a result, King David Kalākaua and his sister,
Princess Liliʻuokalani, watched their influence diminish. When the
time came to renew the trade agreement with the United States,
King Kalākaua insisted that "he would never ratify the treaty and
that the natives would be all opposed to it."

But weeks later, a league of foreign interests, led by Americans,
demanded Kalākaua sign a new constitution that transferred more
power away from the monarchy. Princess Liliʻuokalani reported
that the operatives forcing these changes on her brother had "taken
measures to have him assassinated if he refused," and they had
"prepared measures to be a law unto themselves." The so-called
bayonet constitution named for "malcontents obtaining, by means
of bayonets, a Constitution to suit themselves" set the volatile
backdrop for the *Wandering Minstrel*'s arrival.

Frederick did not want to sail all the way into the harbor, which
would have further slowed them down, but anchoring near Honolulu provided a respite from conditions at sea. Still, this did not
assuage a small number of disgruntled crew members vexed by the
rough passage so far. Sailors routinely endured harsh conditions
with little remuneration, alongside shipmates with uneven expertise
who were prone to misunderstanding each other's languages and

cultures. Some of them came from desperate circumstances or had erratic personalities. When faced with inevitable challenges, breakdowns in discipline and hierarchy spread. The Walker boys and other children who grew up sailing observed the highs and lows of the sailors' choices, an unofficial part of their education. "I am so glad my mother used to give me such good counsel," said another boy who sailed in the era, "for I don't think I shall grow to be so wicked as some of our [sailors]."

The first and second mates blamed Frederick for the punishing schedule and decided they wanted control of the *Minstrel* for themselves. Word of the looming mutiny reached Commissioner James Hay Wodehouse, sixty-three, the primary British representative on the island. Wodehouse had acted on behalf of British interests in a number of locations around the world before receiving his appointment to Honolulu. Safe travel, shipping, and commerce at sea were crucial for British relations with Hawaii, particularly as Wodehouse monitored how the actions of the United States diminished Hawaiian sovereignty. The *Minstrel* fell into Wodehouse's jurisdiction because it was based out of British-colony Hong Kong.

Wodehouse sent a local police deputy marshal named John Pahia on board the *Minstrel* to investigate the possible uprising. Deputy Marshal Pahia was a man known for jumping into action. A month prior, when a carriage lost control in Honolulu, sending the horse running through the middle of busy King Street, Pahia chased and caught it. The police also had skills balancing the need to protect commerce with respect for the monarchy, as when Pahia investigated a theft from a merchant and found the culprit in the residence of a member of the royal family.

Bracing for a physical fight on the *Minstrel*, Pahia chose his most intimidating officers to board the schooner. He was almost disappointed when they did not meet much resistance and arrested the two ringleaders without incident. Frederick immediately put the word out that he needed replacements, and two experienced sailors who knew each other in Honolulu, John Cameron and George Hanker, came on board as first and second mate, respectively.

With the crew filled out, the delay proved brief, and the vessel sailed away from Honolulu on December 10, 1887. After the acrimonious departure of the previous officers, Frederick had to reassure the rest of the crew their positions were safe, including his cook, a hardheaded American named Frank Lord. Like many shipboard cooks, Frank was hardly an epicure, but could claim the faint praise that he cooked better than he sailed. The Walkers' favorite meal was a soup consisting of rice, salted pork, fried steak (turtle or chicken), corn, snap peas, potatoes, pieces of biscuit and fresh fish, with Worcestershire sauce for added flavor. After meals, Frederick would smoke his pipe.

In a week they reached the French Frigate Shoals, a crescent-shape reef on top of a sunken volcano (where, without knowing it, Elizabeth became the first woman to be documented visiting there). While there, they hunted and explored its islands, but at sea the routine of the *Minstrel* lent itself to monotony and existential loneliness. As another sailing wife recorded in a journal of her voyages, "occasionally we catch a glimpse of a distant vessel, and as we are going different directions, they soon vanish, and we are left alone, a solitary speck on the ocean." A glimpse of another sail would be even rarer as the Walkers' voyage through the Pacific continued.

Along the way, Frederick Walker had to build chemistry and trust with his new officers.

The new second mate, George Hanker, fifty-eight, was a battle-scarred fixture on vessels of all kinds. A reporter who visited with him described him as "an iron will encased in a setting of pride as hard as steel," while a traveler who encountered him described him as "a strange looking man." Unlike Frederick Walker's close-knit family—George watched them sing together around the family piano on the schooner—George was decidedly solitary. Decades earlier, he had left behind a young wife and two children, not returning for eight years. When he did show up at home, he remained just long enough to learn his wife had died before he shipped off again. He intended to send back money to support the children, but whatever he'd earn on his long voyages tended to disappear quickly. When that happened, George, taking advantage of his powerful physique, would find work clearing forests with a "reckless" axe.

George was a subject of curiosity, but the other new hire on the *Wandering Minstrel*, the first mate John Cameron, would prove to be transformative. In fact, the fate of the Walkers would pivot upon his actions and his hunger for power.

Chapter 2

THE FIRST MATE'S STORY

WHEN HE FILLED THE FIRST-MATE vacancy on the *Wandering Minstrel*, John Cameron, thirty-seven, was looking to return to prominence.

Writer Fanny Stevenson described John as having a "twinkling, hard blue eye." Quick thinking and opportunistic by nature, he cranked up his silver-tongued charm with ease. When he did, he could be very convincing, deflecting responsibilities and fabricating explanations for questionable behavior at a moment's notice.

Born in the Scottish Highlands, John apprenticed on a ship when he was twelve years old, then spent his teens and young adulthood as part of crews throughout the Atlantic and Indian Oceans. As a sailor in his early twenties, John had been on a coal bark (a vessel characterized by having three or more masts and by how the sails are rigged), *Florence Nightingale*, which sank, preserving only "the clothes worn by the crew" by the time they were rescued by a Spanish steamship.

A formative turn of events in John's life occurred seven years before he joined the *Minstrel*. In 1880, the then thirty-year-old served as second mate on the *Pomare*, a brig that was engaged in "blackbirding," or forcibly recruiting Pacific Islanders as indentured laborers, commonly for Hawaii sugar plantations, a demand driven by treaties with the United States.

The annals of blackbirding overflowed with tales of infamy, but if there was one model of ruthlessness for John to emulate, it had to be James Patrick Murray, an Australian surgeon. Dr. Murray secured a brig called *Carl* and brought a crew to the New Hebrides (now Vanuatu) islands. Concerned about resistance among the islanders, Murray and other crew members dressed as missionaries, complete with Bibles in hand, coaxing more than a dozen islanders onto the ship, where they were locked up below deck. The blackbirders sailed for another island, where they were too impatient to repeat the missionary ruse. Instead, they sank islanders' canoes with iron slabs, so that the islanders in the water could be dragged aboard the *Carl*. When the prisoners rebelled, Murray and crew fired guns blindly into the hold, massacring fifty and wounding seventy more.

John Cameron likewise used violence, trickery, and kidnapping in his attempts to round up laborers, many of whom would never again see their families. *Pomare* captain A. N. Tripp instructed a half-Fijian member of the crew named James Cadigan to row smaller boats with John to the islands, with Cadigan following orders to pose as the captain in an attempt to use his heritage to win the trust of islanders. When they returned from the islands, John would call recruits "our catch for the day."

In a sinister inversion of the Walker boys excitedly boarding ships with their parents, John and his fellow blackbirders would bring children aboard their vessels by deception and force. A journalist who sailed undercover on a blackbirding voyage chronicled families being split apart: "Two or three times when a boy had come aboard . . . we would see from the decks the mother standing on the sand, holding her arms out toward the vessel, bewailing with the agony of a broken heart, clamoring to be taken aboard, rather than lose her child, but held back by her husband and friends, who were powerless to rescue the boy, but who could prevent her going to a similar fate. These scenes were enacted over and over again at every island."

John attributed any resistance by an island's native community to their savagery and responded without mercy. "My one thought was to kill," he said when recounting an incident in which islanders jumped in the water in order to stop his recruiting boat while it was rowing to shore. John fired his rifle at the swimmers. "What I accomplished I do not know, except that I must have accounted for"—by which he meant he shot and possibly killed—"some of the black devils that were falling about us."

At lush, mountainous Pentecost Island, John praised the population as top candidates to be laborers because of their "splendid physique[s]," which was not an uncommon attitude among westerners at the time. Once again, the islanders tried to defend themselves against the *Pomare*, launching canoes filled with warriors. Captain Tripp ordered John to fire the cannon above the heads of the rowers in order to scare them away. Instead, John fired at the canoes. "I'm very sorry, sir," John later apologized to the captain for the cannon's malfunction. Tripp believed his second mate was remorseful, but there was in fact no remorse—and no malfunction. As he later admitted, John gleefully aimed the cannon directly at the canoes, and had even loaded the weapon with a more potent charge of powder.

After his lucrative career as a blackbirder, John was given the opportunity to command a vessel for the Pacific-based Inter Island Steamship Company. He seized it. Rigged with sails but also possessing engines and propellers, the company's steamers were built significantly larger than schooners to include room for cargo and passengers. Their routes and schedules were more predictable than vessels exclusively dependent on sails and wind power.

Among the Inter Island steamships assigned to John was the impressive *Planter*, used for transporting coal and other freight as well as conducting passengers between two Hawaiian bays. According to one captain of the region, it was "one of the finest steamers that ever sailed out of Honolulu."

By then in his midthirties, John was determined to prove himself. But on one passage of the *Planter* with John at the helm, on the foggy night of January 28, 1886, only six hundred feet from an anchorage at the western Hawaiian island Niihau, the bow of the *Planter* collided with the reef. A passenger, Florence Barbour, twenty, slept in her berth. Florence had been struggling with her health at her home in Northern California and had heard that the Hawaiian climate could benefit her. When the *Planter* struck the reef, Florence fell from bed with such force that she suffered a pulmonary hemorrhage.

Everyone on board, including the enfeebled Florence, had to take lifeboats to Niihau. To make matters worse for the distressed captain, after waiting for everyone else to be lowered onto the small boats, a wave pulled John into the ocean. He nearly drowned but was rescued by his crew from the water in an unconscious state, heavily bruised and, after regaining his senses, humiliated.

The *Planter* broke in two, "one half having gone one way," read the report, "and the other half another." The boiler and the smokestack both ended up in the ocean, floating back and forth with the tide. An industrious salvager promptly purchased the rights to recover pieces of the wreck, planning to use the once admired steamship for firewood. A cat who caught rats on the steamer was rescued after five weeks of being trapped in the wreckage. In the meantime, a schooner weighed anchor to rush the seriously injured Florence to Honolulu, amid concerns that she would not survive the trip.

Kingdom leaders at the time wanted people outside Hawaii to know about the "many attractions presented by life in the tropics." Estimates showed 2,040 visitors came to Hawaii the year of the *Planter*'s wreck, but subtracting sailing crew and other "accidental visitors" who were onshore briefly, tourists numbered 600 to 700 per year. This was "a mere fraction of the crowds who flock to the far less attractive resorts of California," in the assessment of those advising Hawaiian efforts. They wanted to capitalize on the attention generated by prominent visitors such as novelist and humorist Mark Twain, who commended Hawaii's "garlanded crags, its leaping cascades, its plumy palms drowsing by the shore, its remote summits floating like islands above the cloud wrack," which comprised "the loveliest fleet of islands anchored in any ocean." In the plans to tempt travelers, safe and reliable means of transportation recurred in the list of priorities: "the times and methods of sightseeing; the means of communication to various parts of the Islands; the interisland steam coasting service as contrasted with what supplied its place in the early days; the excellence of the Oceanic steamships." Oceanic was one of the lines that brought cargo and passengers back and forth between the islands and other countries, along with the Pacific Mail Steamship Company and the Inter Island Steamship Company that had assigned John to the *Planter*.

The crash of *Planter*, and the injury to Florence Barbour, undermined Hawaiian ambitions for public confidence in what was meant to be "a pleasure trip in every sense of the word." "The sea-scented breezes are health-giving and refreshing," promised a contemporary guidebook. Florence, seeking those restorative influences of the tropics, was the type of visitor they wanted to attract, and now she fought for her life.

Once Florence arrived in Honolulu, the responsibility for her treatment fell to Boston University Medical School–trained George

Henry Martin, a clean-cut thirty-six-year-old American. When Martin was a child, he witnessed his father, who served as part of the Eighth Massachusetts Regiment in the Civil War, return home in ill health after being held as a prisoner of war at Belle Isle in Virginia. This experience, combined with an extended stay with a physician uncle, planted early seeds of interest. After his formal training, he became a surgeon overseeing two different soldiers' hospitals, then joined his uncle's practice in San Francisco before taking over a vacant medical office in Honolulu.

In addition to Florence's condition, external pressures mounted on the attending doctor. Dr. Martin represented the current face of the medical establishment of the kingdom, indirectly reflecting its promises to replenish the health of its visitors. On top of this, the patient herself could be said to come from medical aristocracy. She was part of a prominent family of physicians in California that included a doctor who treated Hawaii's princess, Lili'uokalani.

Dr. Martin tended to Florence for weeks. He tried a wide range of treatments that included the controversial methods of Samuel Hahnemann, considered a pioneer in homeopathy. But Florence died on February 28. Honolulu mourned the tragic fate of their visitor. The Young People's Christian Temperance Union, which positioned itself as a moral authority for society, postponed a meeting in order to honor her loss. The tale of the young woman from an influential family who had looked for a better life, "a stranger in a strange land," only to end up in a coffin captivated Honolulu high society and further disgraced John Cameron. An impressive array of flowers filled the funeral, including a "pillow" of heliotrope arranged to display a letter "F" in the middle. A hymn was sung to one of Florence's favorite tunes, with poignant allusions to the sea:

Jesus, lover of my soul,
Let me to Thy bosom fly,
While the nearer waters roll,
While the tempest still is high.

Showing an adroitness in deflecting blame that he would later use as first mate of the *Wandering Minstrel*, John pointed fingers at the Inter Island Steamship Company by claiming faulty instruments and insufficient crew. But with Florence's death fresh in the public consciousness, the odds of securing another post as captain of a vessel dwindled. Instead, John partnered with another former Inter Island captain to try to compete with their ex-employer, leveraging their capital to buy a small fleet of cut-rate schooners for a misguided territorial war against powerful steamships. One of the schooners the new partners eyed was called *Emma*. The *Emma* soon burned in a fire described by the Honolulu *Evening Bulletin* as "mysterious," after which John and his partner were able to afford to purchase the vessel. Meanwhile, needing cargo to turn into cash, John convinced Captain J. F. Luttrell, a fellow "blackbirder," to advance him goods for trading in the amount of $1,600.

John's partner soon abandoned the project, and Inter Island strategically blocked him from becoming a serious competitor. John skipped out on paying his debts, including what he owed Luttrell. Still unable to give up his grudges, John secured work with a rival of Inter Island. While at the helm of that vessel, he might have attached ominous meaning to the volcano eruption he watched from Hilo Bay, with its "flames of great brilliancy." That same voyage marked another unimpressive showing for John, having to steam out of the harbor twice in view of amused onlookers on shore.

With another venture stymied but the fires of his ambition still burning, John's status again slid down to a ship's officer, which brought him to the *Wandering Minstrel* when it had a sudden opening. John had a feeling of big things to come, a feeling that mirrored the Walkers' own. The new first mate had an impression that "this vessel was to take me into strange waters and singular adventures." He was more right than he could know.

Chapter 3

THE SKIES DARKEN

The New Year, 1888

AS THE *WANDERING MINSTREL* CONTINUED on course for the first week after setting sail from Honolulu, Frederick aimed to be equal parts mentor and leader, prioritizing a harmonious operation during lulls and quick action during shark sightings. At French Frigate Shoals, stretches of sandy island formations filled with monk seals, which in turn attracted sharks. One of these formations was later named Shark Island for the sharks known to circle it. The Pacific's coral reefs and underwater volcanoes created navigational challenges that discouraged human presence, rendering an ecosystem uniquely dominated by apex predators. At times, the Walker boys could stand at the bulwarks and marvel at "sharks by the thousand."

According to the superstitions of seafaring people of the time, sharks following or coming near a ship was considered ominous, "a bad sign—in fact, a sign of death," as the Honolulu *Advertiser* put it. Sharking required dismissing these superstitions, since the fishermen needed the sharks to be near their ships in order to catch them. It also required stomaching the more obvious danger posed by sharks, on the one hand, and, on the other, tolerating the gruesome process of killing the sharks in order to sell their fins and oil from their livers. For the Walkers, it meant sharing those unpleasant aspects of the vocation with their children.

Some shark fishermen matched whatever violence sharks might display: sometimes those fishermen manifested sheer barbarism, fueled by the notion that "of all the enemies with which man has to contend, there is not one more to be dreaded than the Shark." Not long before the *Minstrel*'s voyage, the crew of a vessel based in Gloucester, Massachusetts, cut off a shark's head and threw the body back in the water, watching as the headless body reportedly swam for forty-five minutes—a story picked up by newspapers worldwide, contributing to a perception of sharks as otherworldly, even demonic, in an era when most people had never seen one. As late as the sixteenth century, no word existed in the English language for shark. Many people who were later ridiculed for their detailed reports of sea monsters had been describing very real sightings of sharks. "We refer to a monster that no one has seen," read a late nineteenth-century article on sharks, "except those who have visited that portion of the globe where the Shark is at home."

John Cameron believed Frederick lacked the ruthlessness to be an effective sharking captain. More broadly, the new first mate saw in his superior on the *Minstrel* a leadership approach that clashed with his own. He thought Frederick was simultaneously too aloof a commander and too easy on his underlings. Charlie Walker saw how the first mate sneered at his family, and noticed that John treated him and his two brothers like children, whereas fourteen-year-old Charlie was determined to prove himself a man.

Conditions at sea kept Frederick occupied, requiring him to change course several times to avoid storms. By late January 1888, those aboard the *Minstrel* believed they had entered a historic season for storms in the North Pacific. It seemed to Frederick as though the bad weather followed them. They felt like they were sailing into the middle of nowhere, where appropriately enough they spied the Midway Atoll. The name reflected a location that was an approximate midpoint between North America and Asia. Midway was also the northernmost atoll, or geological formation of coral, in the world, and one of the most isolated pieces of land on earth. When Charles Darwin voyaged on HMS *Beagle* fifty years earlier, tasked by the Royal Navy to study atolls, Midway would have made an excellent setting for investigation, but the ship's navigators did not find it. Robert Louis Stevenson and Lloyd Osbourne described the daunting approach toward the leeward side of Midway (that is, the side sheltered from the wind) in *The Wrecker*, which would have mirrored what the Walkers and their crew saw: "We had gained the lee of the island as (for form's sake) I may call that ring of foam and haze and thunder; . . . I climbed into the rigging, stood on the board, and eagerly scanned that ring of coral reef and bursting breaker, and the blue lagoon which they enclosed."

The few people of the era who had been to the atoll, formed by volcanic activity millions of years earlier, spoke of strange qualities emanating from the place. "Desolation of desolation," one later visitor wrote in a list of phrases with which he tried to express the atoll's atmosphere, "weirdness intensified, the extreme of melancholy."

The *Minstrel* anchored in a bay known as Welles Harbor, from which they could observe the islands and begin to land their fishing equipment as weather permitted. With sharks swimming around Midway's reef and two islands on a regular basis to hunt, they decided to wait. But the barometer plummeted, indicating more storms coming.

THE SKIES DARKENED the first day of February 1888, battering the *Wandering Minstrel* with rain.

For three days and nights, the ship was pummeled by typhoon-force "gale after gale." It was during these storms that an anchor broke and the *Minstrel* tumbled against the reef, sending the Walkers preparing the lifeboats and desperately searching for their dog. Down in the water, the sharks that they had hoped would be a boon to their livelihood now turned into terrifying threats.

The Walkers could not help thinking about a similar scenario that occurred two years earlier on a schooner called *Pohoiki*, especially because that ship's captain also brought his wife and child. When the *Pohoiki* began sinking off the Kau region of Hawaii, the captain and his family and all-Hawaiian crew made it into lifeboats, but the lifeboats capsized from the suction of the submerged schooner. While struggling in the water, the captain's legs were bitten off by a shark, while another crew member had his arm torn off. As a survivor later recounted, "the men . . . were shrieking as the sharks pulled them to their death." The captain drowned, and his wife and child disappeared underwater, presumably also taken by sharks.

The Walkers, facing their own life-or-death scenario, layered on clothes for warmth while trying to ensure nobody on board would be left behind.

Down below in the hold of the *Minstrel*, John Cameron's cruel nature emerged. He rushed around among scared crew members who raced to save themselves. John eyed the storeroom, then stopped a young Chinese "lad" named Mouwen, whom John called Moses. John ordered him to go into the storeroom to retrieve tobacco. Mouwen, frightened, tried to do as he was told, but the storeroom flooded. Mouwen insisted that if he went inside, he could die. John replied threateningly that Mouwen might also die if he did not go in. Knowing that the errand became deadlier every second, Mouwen went inside and came out with a box of tobacco for John. The first mate grabbed it, remarking, "You velly smart fellow, Moses," in imitation of Mouwen's English.

John and Mouwen made it to the deck. But as everyone aboard took turns lowering themselves down on a line, John was afraid to try to get into the lifeboats that waited for them in the rough seas. The irony was rich—in the lower deck, he had selfishly forced someone else to put his life at risk, while up above he was too scared for his own safety to follow orders. Every minute that the first mate delayed put the entire group in more danger. On sailing vessels in crisis, even a brief lapse in protocol could initiate a chain reaction of problems. Elizabeth Walker understood this. Her nephew, Richard Owen Eyton, acted as chief steward of a vessel called *Marloo* when a crew member died after falling through an open hatch. Other officers initially blamed Richard for not following orders, dragging him through a legal process before his exoneration.

On the rapidly foundering *Minstrel*, Elizabeth and the younger Walkers elicited a promise from John that if they demonstrated that they could reach the boats, he would consent to follow. After the three Walker boys made it in safely, and Elizabeth just barely landed in the boat, they heard Jessie's bark.

The dog appeared above them on the deck, running in circles and yelping. Elizabeth pleaded with the sailors still on board to rescue their adored pet, whom they considered part of the family. One sailor, sick of the noise, picked Jessie up and carelessly tossed her overboard. By good aim or good luck, Jessie landed right in Elizabeth's arms, and she cradled her safely into the lifeboat.

Breaking protocol, sailors piled personal possessions into the three boats until Frederick demanded they stop. In the back-and-forth, a bag of biscuits fell into the sea, leaving them without food. But they'd worry about that later. The Walkers made sure every crew member, including John, was on a lifeboat. They had to push away from the sinking *Minstrel* to avoid the notorious fate of the *Pohoiki*'s family. From their new vantage point, conditions were all the more frightful. "No man could have said where air left off and water began." They prayed as they rowed the small vessels into the storm, with nothing but oars to protect themselves.

Chapter 4

THE EDGE OF THE WORLD

February 4, 1888

THE CASTAWAYS AWOKE THE NEXT morning on land. Their lifeboats had been able to reach Sand Island, an arid formation lodged beyond the reefs at the Midway Atoll. "Island" was generous: Sand Island was closer to being what observers called a "barren sand bank" or a "sand heap of irregular shape." Frederick's second mate George Hanker, after exploring, judged it "little more than a sand dune on the bosom of the Pacific." It ran a mile and a half in length, three quarters of a mile wide and a mere fifty-seven feet above the sea at its highest elevation, which meant the shape of the island changed with the water levels. As indicated by its simple name given by surveyors twenty years earlier, the island was almost entirely sand, which in the sun produced a blinding whiteness, and with very little in the way of vegetation or fertile soil. Across a channel from Sand Island was an even smaller stretch of land known as Green Island, which was just as unwelcoming.

Steering the small lifeboats the previous evening had been a torturous, exhausting experience, with one of the boats capsizing and temporarily trapping Frank Lord, their cook. As the prospect of drowning flashed over him before rescue, Frank could put the blame on one person for his near death: first mate John Cameron, whose delays in casting off the lifeboats had allowed the breakers to grow.

They only realized now, in daylight, what a remarkable feat it had been to have reached land. From the safety of shore, they could study the giant waves and ragged coral formations they had been up against. "I shuddered," John admitted, "as I gazed over the smoking sea through which we had passed." As the Walkers looked out from the beach they could make out what remained of the *Minstrel*, most of which had sunk while some wreckage floated. There were thoughts of taking a lifeboat out to where they saw wreckage to search for provisions. But storms continued to rage, and soon the foremast (the mast closest to the front of the vessel) and the mainmast (the tallest one) crashed down into the water, dragging more wreckage with it below the waves. All that was left in sight was "a portion of the stern with the mizzenmast laying upon it," that final mast seeming like "a monument to a vessel dispatched to her doom." The vessel with which Frederick had envisioned achieving financial stability for his family was now a lost cause, gradually smashed by waves into scraps swept away to what might as well have been the ends of the earth.

On the positive side, everyone had survived—the Walkers and twenty-four crew members. The castaways spread out to try to identify wreckage that was closer to shore. Part of the stern (or rear of the ship) still floated at sea, likely caught on the reef. They were able to wade out and extract what they could. They found a few pieces of clothing and tins of provisions, the latter being a cause for celebration.

A few sails and some awning material were also retrieved, as were fragments of the family's piano, a reminder of domestic scenes starkly different from their current setting, particularly for Elizabeth, for whom music always accompanied family life.

Freddy, Charlie, and Henry helped to gather and organize the remnants. The boys' transformed circumstances contrasted with the excitement they felt when their voyage had begun, when they had explored the vessel's dramatic armaments. Now the ancient cannon that had fascinated them sat at the bottom of the sea. Once-imagined battles with pirates would have been less daunting than being stranded on Sand Island. Their father did manage to bring to shore some smaller firearms, his revolver among them.

Older sailors among the crew distracted shipmates by gabbing about disasters from their pasts. For hulking second mate George Hanker, a favorite in his repertoire involved finding himself in the Australian Outback, describing how he "lived with naked savages, subsisting on ants, snakes, cockatoos and parrots."

Despite all the Walkers' personal experiences at sea, the primary available model for a family of castaways on an island would have been a novel, *The Swiss Family Robinson*, which was actually not a fixed text, but a fluid one that changed titles and content over time. About seventy-five years earlier, the original two-volume version of that novel, a twist on the fictional castaway tale *Robinson Crusoe*, was published in German as a dramatization of how a husband and wife rear their children while stranded on a highly idealized desert island. From 1814 to 1818, an English version prepared and expanded by writer Mary Godwin (stepmother to *Frankenstein* novelist Mary Shelley) was published in two parts as *The Family Robinson Crusoe* and then in a single volume as *The Swiss Family Robinson*, the version that would have reached the Walkers. Uncannily, the fictional ship of the novel, like the *Minstrel*, also wrecked on a coral reef, the fictional family had four sons (to the Walkers' three), two dogs (to the Walkers' one), and the fictional mother's name was also Elizabeth. The Elizabeth of the novel, a background character in the German original, became more prominent in the French and English versions that were translated and modified by women.

The real Elizabeth stranded on Sand Island was understated by nature and self-sufficient, qualities deeply challenged by her new environment. She did not seek out attention and never wanted to make herself conspicuous, yet now found herself at the center of the unprecedented real-life *Swiss Family Robinson*.

They had to assess the odds that a ship might come by and spot them, given that the Walkers and their crew knew Midway's status as one of the most isolated—arguably the most isolated—pieces of land on the globe. Midway was 3,700 nautical miles east of Hong Kong and 2,200 miles east of Tokyo; it was 2,700 miles west of San Francisco and 1,150 miles northwest of Honolulu; going south from Midway would require traveling 3,750 miles before reaching Australia, and almost 3,900 miles to New Zealand. For all practical purposes, they found themselves at the edge of the world.

The first person documented to have stepped foot on Sand Island had arrived twenty-nine years earlier. That sailor was an American named N. C. Brooks. The atoll had formed millions of years earlier, but Brooks found no sign of anyone ever having been there, although he spotted a raft nearby that he speculated had been brought by the currents. He later publicly suggested the islands be named after him, and also proposed that the atoll be used for strategic and commercial purposes, which inspired the United States government to arrange another expedition nearly a decade later in 1867 to claim the atoll. The leader of that expedition, Captain William Reynolds, embraced the ceremonial aspect, announcing afterward that "I ventured to name the only harbor at [Sand] island after the present Honorable Secretary of the Navy [Gideon Welles]." Reynolds described the pomp of the expedition, how he "landed that day, accompanied by all the officers who could be spared from the ship, with six boats armed and equipped, and under a salute of 21 guns, and with three cheers, hoisted the national ensign, and called on all hands to witness the act of taking possession in the name of the United States." He also viewed his mission as the start of an era for the United States: "It is exceedingly gratifying to me to have been thus concerned in taking possession of the first island ever added to the dominion of the United States beyond our own shores, and I sincerely hope that this instance will by no means be the last of our insular annexations." That set up yet another voyage and plan, an "ill-starred expedition" that ended with the deaths of four men.

THAT DOOMED MISSION of 1870, three years after Captain Reynolds's ceremony and eighteen years prior to the *Wandering Minstrel*'s arrival, intended to transform Midway, and was championed by the Pacific Mail Steamship Company. The plan entailed using gunpowder to blast through the massive ring of coral that surrounded Sand Island, creating a wide channel where vessels could anchor at the island, on which the United States would construct a coal station. The station would allow steamships to refill coal used to fuel their engine boilers during long passages.

Naval officials chose USS *Saginaw*, based at the Mare Island Naval Shipyard in Northern California, to lead the mission to Midway. The *Saginaw* was a sidewheel steamer, with a wooden hull containing iron machinery and engines that turned paddlewheels on both sides of the ship in order to generate movement without relying on sails. Under Lieutenant Commander Montgomery Sicard, the *Saginaw* transported an engineer, contractors, and supplies from the San Francisco area to Midway. As they approached the atoll, the supply officer, George Read, described the sight of Sand Island, where the Walkers would later be stranded: "We can see the white sand drifting about with the wind like snow." In less picturesque moments, the sand thrown forcibly by the wind cut their faces, necessitating a supply of goggles. Even a crew trained by the United States Navy dreaded the surrounding reef. With a strong gale that swept in, even "with both anchors down the engines had to be worked constantly to prevent drifting either on the island or the reef."

The *Saginaw*'s mission directly impacted what the *Minstrel* would later find there. The *Saginaw* crew built a small house—or hut—with a door to provide them with shelter from sand and rain. Meanwhile, the contractors also attempted to blast through coral to widen the passage. They sent a diver in protective armor down through the reef, where he placed tin canisters of blasting powder into the crevices of the coral. These were triggered by electricity. The contractors aimed for each explosion to dislodge a couple of tons of rock. Over the course of several weeks, they built a heavy scow—or wide-bottomed open boat—to push out to sea over the reef where the diver was at work. On the scow was a derrick (or crane) attached to an engine. The diver underwater would place the pieces of coral broken off by the blasts onto the hoisting gear, which would be lifted up by the device into the scow.

But storms and gales disrupted them. After six months of their arduous process, the passage through the reef had been widened to only 15 feet, far short of the 170 feet that was called for. Yet they burned through their $50,000 appropriation.

On trips to Honolulu to pick up supplies, Captain Sicard sent discouraging reports back to naval leaders, with the ultimate conclusion that they would need several years and four times the original appropriation to make Midway viable. Moreover, Sicard had come to believe Midway's harbor would never be safe for large steamers operated by Pacific Mail, even if they reached their goals. The mission was deemed a lost cause, and the *Saginaw*'s officers prepared to take everyone off Sand Island.

Naval personnel and contractors alike were glad to leave. The ship's surgeon loitered before boarding the *Saginaw*, insisting that he felt the ship would meet disaster, as though cursed from their time at Midway. His concerns were dismissed. Since they could not fit everything in the *Saginaw*, they abandoned the scow they had constructed, pulling it high up on shore. Once under way at sea, Captain Sicard decided they would take a fifty-mile detour to Ocean Island, which was really two islands surrounded by dangerous barrier reefs. He wanted to survey the reefs at the uninhabited formations (later called Kure Atoll) to ensure accurate details on future charts. Considering that Ocean Island's history included some of the earliest recorded shipwrecks of Hawaiian vessels, Sicard searched for any castaways. The captain later emphasized this motive: "My errand to this island was purely one of humanity, I [was] not feeling willing to leave these parts without seeing if there might be some shipwrecked persons on the island, whom I might be able to relieve." A survey to prevent shipwrecks or even a hypothetical rescue could also compensate for the underwhelming results of their official mission to Midway, making the outcome ironic.

Approaching Ocean Island in the dark, the *Saginaw* struck the reef, which tore through the hull and engine room. The dangerous reef that Sicard had wanted to document had claimed their vessel. The crew found their way onto land in lifeboats while the steamer broke apart. They set up an encampment on the highest possible ground, where they organized the food and supplies they had managed to salvage. Storms plagued them, and they decided to send an open boat rigged with sails to try to reach Hawaii (then known to Americans as the Sandwich Islands). As five sailors pushed off in the twenty-two-foot lifeboat, the other castaways prayed, and one of the older sailors opened his Bible by chance to Isaiah 51:5: "the isles shall wait upon me, and on mine arm shall they trust"—meaning even those on the most distant lands would find strong faith in God.

The crew on the lifeboat sailed for thirty-one days, with little or nothing to eat for days at a time, traveling a total of fifteen hundred miles. Finally in sight of the land, heavy breakers tossed the boat, washing overboard three of the men, who drowned. The brawny coxswain, William Halford, twenty-nine, reached land, where the only other survivor collapsed and died. Hardly able to walk, Halford came upon a group of young islanders, who ran away. "I must have been a sight to behold!" Halford later recalled. "A couple of months' beard, long hair, dirty, naked, and wasted almost to a skeleton!" Halford discovered he was in Hanalei harbor on the Hawaiian island of Kauai. Transported to Honolulu, Halford reported the location of the *Saginaw* castaways to the American consul. Ten days later, the rest of the castaways on Ocean Island were rescued. Captain Sicard later recalled seeing the ship "as soon as she appeared on the horizon, and our feelings of gratitude may be imagined, perhaps, but can only be thoroughly appreciated by those who have been placed in a like situation to our own." To the probable horror of the ship's surgeon, who had felt vexed by Midway, the steamship that took the castaways off Ocean Island stopped back at Midway's Sand Island in order to fuel up with coal. On finally reaching Honolulu, "the esplanade presented a scene seldom witnessed," observed the captain of the rescue vessel. "As she neared her dock, the crowd, unable to control their feelings at the sight of the weather-beaten faces on board, gave vent to hearty cheers, which were returned lustily from the steamer."

Though the wreck had occurred at Ocean Island, the story of the *Saginaw* was inseparable from Midway. In fact, at least one mention in the press merged the events, referring to the *Saginaw* as having wrecked at the Midway Atoll. It was as though Midway had doomed the ship, and in turn led to the U.S. government giving up plans for the atoll. As one columnist in Hawaii quipped of Midway, "Uncle Sam left his claim to the original owners, the birds and crabs."

TAKING INTO ACCOUNT the expedition of the *Saginaw*, the Walkers' best estimates indicated that a ship came to Midway approximately once every five years. They had to face the fact that they would likely be on their own for quite a while. No matter how industrious, Elizabeth and her family now depended upon ingenuity and many external factors just to survive—including the weather and the morale of their shipmates. The most urgent needs were food and drinkable water. Unlike the circumstance of a vessel called *Henry James*, which had wrecked around the same time as the *Minstrel* more than a thousand miles to the southeast at Palmyra Island, which had coconuts that could be consumed, Midway presented few natural resources. Still, while the *Minstrel* castaways found almost nothing edible growing on Sand Island, there were signs that turtles and seals occasionally came there. There were also several species of birds, including the southern albatross, called goonies for their clumsy, awkward movements when landing or walking, in contrast to their graceful flight. They were loud and boisterous enough that one moniker for Midway in Hawaiian became *Pihemanu*, which meant "the loud din of birds."

John Cameron noted that what he called mutton birds (wedge-tailed shearwaters), another possible source of food, created particularly disturbing noises. "The first time I heard one crying I received a sharp shock of fright, so mournful, uncanny, unearthly were its wails as it sallied from its underground nest to search for food in the gloaming."

The castaways had to find ways to rig tools with which to catch fish. Frederick always cultivated an inventor's approach to challenges, a passion embraced by his children, particularly his youngest son, Charlie.

While they experimented with creative approaches to fishing, the eggs of the island's birds became the primary ingredient in almost every meal they had, in lighter moods given names like "egg tea" and "egg pancakes." In addition to the eggs, the castaways would catch the birds, clean them as thoroughly as possible, and cook them over fire.

Frederick felt ill without tobacco, which he was accustomed to smoking in his pipe after his meals. (He had once tried to give up tobacco for a single voyage, not permitting it on board, but out of desperation he and his crew tried to smoke chamomile leaves. "I never again tried to do without smoking," Frederick later declared.) If the lack of tobacco on Sand Island ranked far down the list of problems, Frederick and others felt it acutely, physically and mentally.

John Cameron did not mention his own supply of tobacco, which he had coerced Mouwen into retrieving from the ship. It was not the only thing John wanted to keep quiet. He would not bring up his ignominious part in the notorious wreck of the *Planter*, the accident that fatally injured Florence Barbour. He also had no desire to rehash his unwillingness to rappel into the *Minstrel*'s lifeboat, a delay that endangered his shipmates. His career would not be able to withstand another blow to his reputation.

On Sand Island, John immediately took a hard turn against the crew. "I know these cattle better than you do," he said to Frederick, urging him to establish a kind of martial law. "They must be kept in hand. Dealt with firmly. Taught, especially now, to look up to us as their superiors and masters." The isolation of the atoll, the separation from the rest of the world, presented an opportunity for John to reinvent himself as an uncompromising commander. He requested one of the firearms. "I shall muster the men in line, the first who refuses to obey you will be shot. . . . Do you consent to my proposal?"

"Mr. Cameron," Frederick said, "I do not. That is final."

John resented Frederick's position. He took satisfaction when a squabble among the crew led to a boatswain wielding a spear, which during the ensuing melee sliced Frederick's cheek. The figure of the towering and slender captain, a reassuring sight on the decks of the vessel, now staggered across the beach. With Frederick bleeding, Henry Walker rushed to find John to plead for help. John glared with his steely blue eyes at the young man. "Your father made his own bed, Henry, and there he must lie," he said coldly. He further underlined his position that he was the superior leader: "He should have let me shoot one of them."

An idealist by nature, Frederick, unlike John, believed that if he went by the rules and upheld fair principles, everyone involved in a venture would end up satisfied. However, a sea captain's decisions impacted so many people on board and off—officers, crew, his own family, investors, and trading partners—that those interests rarely aligned, inevitably leading to dissatisfaction and to Frederick feeling demoralized.

After tending to his wound, Frederick managed to restore the peace in the wake of the worrisome skirmish. But the blood stains on the sand combined with John's cutthroat approach signaled unsettling turns to come. *The Wrecker* described violence experienced by castaways on the atoll this way: "The passions of man had sounded . . . and all men trembled at the possibilities of horror." The solitude would be amplified at night, when the seas calmed, when their bonfires ran low, and the only light came from the stars above, as though the castaways floated in the universe. Each *Minstrel* crew member and each member of the Walker family would be tested to maintain focus and resist despair. They would also need to foster open communication and trust with each other. In a barren environment, secrets inevitably would surface.

Chapter 5

STATE, WAR, NAVY

THE FATE OF MIDWAY, AND whether it might be visited in the foreseeable future—and thus whether the Walkers, who nobody knew were there, would be rescued—rested in part upon discussions 5,500 miles away in the heart of Washington, D.C., in the massive State, War, and Navy Building, which might as well have been on a different planet from Sand Island.

Beyond a grand entrance facade, a broad granite staircase with a shiny bronze balustrade led to the second-floor State Department offices, where Secretary of State Thomas F. Bayard educated himself on Midway Atoll.

Bayard, fifty-nine, in his typical outfit of blue-flannel coat and vest with gray pantaloons, oversaw his department with restrained, humorless diligence. A reporter observed him: "Tall and straight, his movements were graceful, and his cleanly shaven face and iron-gray hair were classic in beauty. Broad in intellect, he was patient and courteous in debate, rarely losing his dignity or his temper." Bayard had established himself in politics by representing Delaware in the United States Senate—a seat previously held by his father, grandfather, and great-grandfather—before his former rival, President Grover Cleveland, appointed him secretary of state. Bayard had long weathered criticism for advocating compromise at times when others demanded confrontation (including after the Civil War), and he extended that philosophy to international affairs.

Bayard's State Department was known for providing office jobs to those who were shut out of other work or who wished to study government. When congressional representative and former Pennsylvania governor Andrew Curtin, a staunch Republican who had sat on the stage as Abraham Lincoln delivered the Gettysburg Address, visited Secretary Bayard's office, he became outraged when he thought a young Black messenger did not pay attention to him. Curtin reportedly flew into "a violent rage" while raising his cane over the young man. He screamed: "Stand up this moment and assume a proper attitude . . . if upon another occasion I see such deportment I shall either break this cane over your shoulders or report you to your master." Bayard, even in the halls of his own office, felt he had to be the constant peacemaker.

Midway, an uninhabited atoll in the center of the Pacific that few Americans ever heard of, landed on Secretary Bayard's agenda after becoming a topic in recent diplomatic talks. During the process of renewing the United States' trade treaty with Hawaii, some suggested that Hawaii cede the lagoon of Pearl Harbor to the United States for naval use. With his instinctual stance toward minimal interference, Bayard put forth a counterproposal that Midway Atoll, already legally possessed by the United States, might serve the same purposes for the United States Navy without requiring Hawaii's cessation of rights related to Pearl Harbor.

To make his case, Bayard needed to better understand how Midway came to be claimed by the United States. As a matter of fact, Bayard could learn some of the details about Midway's history walking through the campus of government buildings around D.C. The museum at the Navy Academy in Annapolis,

Maryland, for example, displayed several artifacts from the *Saginaw* that wrecked after leaving Midway, including the lifeboat that brought William Halford to Kauai. One *Saginaw* survivor, upon seeing this lifeboat again, was struck at how "its sad story was almost told in the scars upon it. . . . Its wounds seemed almost as making a mute appeal for sympathy, and expressed the struggle it had gone through." *Saginaw* hero Halford had been celebrated—he received the Medal of Honor, which Bayard could read up on in the government's lists of honorees—but the mission itself remained an embarrassment, a testament to how a tiny speck of an isle had defeated the ambitions of one of the world's great naval powers.

When the time came to meet with advisers to talk about Midway, Secretary Bayard stepped out of his private office into the adjoining reception room. It was considered "the most sumptuous apartment Uncle Sam has yet furnished." A highly polished ebony table and matching furniture filled the room, with a massive fireplace at one end, an ornate iron ceiling, and portraits of past secretaries of state lining the wall, while large southern-facing windows presented a view of the Potomac and the Washington Monument. Bayard would begin meetings by shaking hands with his callers and, without much ceremony, request the latest intelligence.

The full briefing of details on diplomatic talks in Honolulu provided Bayard with more complications to consider. Bayard's suggestion that the United States Navy utilize Midway instead of asking Hawaii to hand over Pearl Harbor raised a thorny question: could the United States continue to authoritatively claim Midway? The United States had left Midway uninhabited for many years. Germany had been declaring protectorates over a number of Pacific islands. Might Germany, or perhaps Great Britain, try to claim Midway for themselves? There were even those in Honolulu who believed that Hawaii had a stronger claim to Midway than the United States. Tussles over rightful possession aside, one possibility, for the time being, was a swap between Hawaii and the United States—Hawaii granting the United States control over Pearl River's harbor while the United States gave Hawaii possession of the Midway Atoll.

But certain details were abruptly resolved, at least for the time being, by the "bayonet constitution" forced upon the Hawaiian king. This ensured the reciprocal trade treaty would be renewed with terms favorable to the United States. At the relevant legislative assembly, King David Kalākaua sat upon the royal chair on a center platform, with his sister, Princess Liliʻuokalani, beside him. Bayard was not present in Honolulu at the time, but others connected to the American government were, including Captain George T. Davis, who commanded USS *Juniata*, one of several United States Navy vessels anchored at Hawaii as part of the Pacific Squadron to protect Hawaii—and the American interests there. Representatives of other nations attended, including British commissioner James Wodehouse and German consul Henry F. Glade. With the treaty enacted, the king also proclaimed the amendment granting the United States a right to develop a harbor at Pearl River.

Bayard doubted his government would act upon that amendment, as the harbor would require significant investment. The provision could also provoke a backlash from those decrying American infringement on Hawaiian autonomy. A prime example was a loosely organized "secret society" that formed in Hawaii, committed to bolstering the royal family against impinging powers.

One faction hatched a plan to remove King Kalākaua and replace him with his more assertive sister.

At the center of this plot to elevate the princess was a thirty-three-year-old named Robert W. Wilcox, the son of a ship captain from the United States and a mother supposedly descended from Hawaiian royalty. A year earlier, Robert, a one-time Maui legislator, had been studying in Italy to be a military leader and strategist, but the bayonet constitution ended his funding from the Hawaiian government. When he returned to Honolulu, Robert, tall and lanky with an eye-catching mustache, was angry with what he considered a puppet government controlled by the United States. His agitation caused problems with his marriage to a woman he had met in Italy. Princess Liliʻuokalani felt compassion for the couple. "In pity for their situation," the princess wrote, "I offered them quarters under my roof until they could provide for themselves." But Robert would not yield his plans. Recruiting approximately three hundred men, Robert stationed them at strategic points and demanded the king abdicate. Unsurprisingly, Robert failed, and had to accept financial support from Liliʻuokalani in order to flee to San Francisco.

Within weeks of Robert Wilcox's departure from Honolulu, Secretary Bayard welcomed two of Honolulu's American business and political leaders, who had come from Hawaii to Washington to share with Bayard their apprehensions about the broader political turmoil. As Bayard wrote in his notes of his meeting with them, they "did not think the King's authority could be long upheld . . . sooner or later the United States might be called upon to assume control of the Islands." There was also a fear "there would be a crumbling away of power, and then no one could foretell what would happen."

Those in the Hawaiian government who favored increasing American power had noted that during these "conspiracies and counter-conspiracies," the flare-ups "were only prevented from breaking into open conflict by the pressures of a fleet of British and American war vessels in the harbor."

That brought Bayard back to the pressing issue of whether the United States needed to cultivate a new naval staging ground in the Pacific. The Senate, drawn to the question, passed a resolution asking for intelligence from the Cleveland administration about Midway. The president in turn requested information from the State Department, and Secretary Bayard compiled a confidential document. With his work piling up, Bayard tried to find time for horseback riding, fishing, and swimming to balance out the monotony of hours spent in meetings and drafting his reports. His wife and one of his daughters had died a year into his tenure as secretary of state, leaving him battling grief while supervising a household of children and young adults. But respites from his office tended to be brief.

As his ongoing series of meetings surrounding the subject of Midway progressed, Bayard consulted Secretary of the Navy William Whitney, whose department was run out of offices below the State Department. Whitney could share more records from the *Saginaw*'s failed mission to Midway. Bayard would stand listening with a tilted head, one hand on a standing globe of the world, and his other hand in his pantaloon pocket.

All the developments pointed to Midway, this little place not even on Bayard's globe, requiring a visit. "It has occurred to me," Bayard concluded, "that it might be as well for the United States to make some expressive movement—by way of a visit of one of our national vessels to Midway Island, the re-erection of a flag and mast and possibly landing some coal there . . . to answer any suggestion of nonuse, or abandonment of title by the United States." Among the other artifacts on display in the government's museums, the navy still possessed water-stained 1870 log books from the *Saginaw*, containing detailed soundings and surveys of Midway that could be used to navigate the treacherous reefs.

They would deploy a naval vessel to Midway for the first time in nearly two decades, which also could be the greatest stroke of luck any group of castaways ever had.

Chapter 6

THE STRANGER

February 1888

THE *WANDERING MINSTREL* CASTAWAYS LOCATED the scow that had been left behind by the *Saginaw* crew. It was no longer seaworthy. Exposed to the sun, sand, and storms for the last seventeen years, the wood had rotted and become brittle. They also came upon the weathered, rickety wooden hut, built at the highest elevation for coal storage by members of the same American expedition, before the environs humbled the United States Navy and the contractors into abandoning the project.

The hut was a crucial discovery. It could provide a semblance of shelter, even if only a few of them would fit inside at a time. But they were soon in for a shock: someone was already living inside. A man in his midtwenties, strongly built but ragged and emaciated, described by a reporter who met him elsewhere as "tall, raw-boned . . . with a large head and lantern jaws," stepped out and raised his hands in a halting form of welcome, seeming "profoundly moved" at the sight of other people.

They could hardly have imagined such felicity; the handsome, light-haired stranger with blue eyes, who introduced himself as

Hans Jorgensen from Denmark, had been stranded there for a while and could educate the castaway family and their crew about the atoll. He explained how he had survived a shipwreck on another island and floated over to Sand Island on a raft. Help from an experienced survivor could be the difference between life and death on the island and a shortcut to finding a way off.

In a gallant gesture, Hans offered use of the hut to the Walker family so Elizabeth and the boys could rest. He shared his ample intelligence. In addition to the scow, planks of wood had been abandoned by the crew of the American expedition that they could now use to extend the hut to create more shelter. Hans confirmed that the limited vegetation and arid soil produced nothing edible. There were the local seabirds to catch and eat, and even more so the bird eggs, which could be collected and consumed. As for fresh water, digging a well, Hans explained, three to four feet deep would provide drinkable, if discolored, water. John Cameron, who had been disdainful of his shipmates and bullied young Mouwen, had an unusual rapport with the new addition.

The Walkers' famous fictional counterparts from *The Swiss Family Robinson* also found a stranger living on their impossibly bountiful island: a young woman named Jenny. Similar to Hans at the moment of his discovery, the Jenny of the novel was at first timid, but came to trust the castaways and to be embraced as part of the family.

As for the dynamics unfolding on desolate Sand Island, there was reason to worry that some of the crew from the *Minstrel* might not prove so hospitable toward Hans. Even if they seemed cooperative at the beginning of their ordeal, some of the sailors could turn on a dime when they realized Hans possessed his own resources, like his collection of birds' eggs. Hans, worn down physically and emotionally, could be vulnerable to becoming a target of violence, and indeed he appeared to worry about a breakdown in behavior from the island's new inhabitants. "Might would be right," Hans reportedly commented in forecasting the prospects of such a relatively large group of castaways, believing that the strong eventually would take whatever they wanted. Elizabeth could empathize; having been the only woman on so many voyages, she knew what it was to feel like an outsider and to always stay attuned to potential hazards.

Frederick, for his part, was a man who thrived on hierarchy, fairness, and order, and was accustomed to being in control. But his rigidness could alternate abruptly with leniency, and one observer thought the captain had fitted his ship "more like a gentleman's yacht" than a sharking vessel, an observation supported by the presence of the family's cherished piano and their propensity to throw holiday parties.

The aborted near mutiny among the *Minstrel* crew at Honolulu in early December now loomed large. While the ringleaders had been taken into custody by Honolulu authorities, there had been other malcontents among the crew whom Frederick had chosen to give another chance. Now those crew members' impulses toward rule breaking could be reawakened, directed toward this stranger on the island, as well as possibly turned against the captain—a rank that could be questioned altogether in the absence of a ship.

The Walkers would have to work together to maintain harmony and use each of their individual strengths to improvise resources. The three Walker sons adjusted. Having spent their childhood sailing meant they'd acquired a wide range of skills, each nurturing his own interests. Freddy, the oldest, was the most artistic and imaginative; middle child Henry was the athletic and competitive thrill seeker; youngest Charlie was a born leader and natural engineer, a valuable trait as the castaways worked to secure food.

Fish, as it turned out, generally did not swim very close to the shores of the island, but during a trial period fashioning tools for fishing, the castaways had begun to rig steel piano wire, recovered from the broken instrument, rubbed with oil to prevent rust, into hooks to reach those that came closest. They could also cut up empty tin cans, once containing provisions, that the *Saginaw* crew had discarded. The castaways unraveled the canvas recovered from the stern of the *Minstrel* wreck to form twine into a net. At one point, two of the sailors wading into the water managed to jam the hooks into a giant albacore, which put up a half-hour struggle before they brought it to the beach. They also used their new nets to catch mullet, ideal for feeding the group because of their thick layers of fat. The castaways relished these meals.

They used the planks left behind on the island to add sections to the hut to create what they called a village of shelters. They decided to break up the *Saginaw*'s abandoned, rotted scow to retrieve the long spike nails for other uses.

The group dug two more wells in addition to one Hans had made. They also collected sea water and allowed it to evaporate so they would be left with salt, which could be used medicinally and to preserve any excess food they caught. They prepared for a long stay.

WHILE THE CASTAWAYS organized themselves, back in Honolulu the United States Navy steam sloop *Juniata* departed on a voyage to Yokohama, Japan, under the command of George Davis, who had attended the legislative assembly marking America's controversial increase in control over many Hawaiian policies. As requested by Secretary of State Bayard, Secretary of the Navy William Whitney sent an order to Honolulu instructing Commander Davis to stop on the way to Japan at the Midway islands in order to "make a report on the state of affairs there" almost twenty years after the atoll had been claimed by America.

A Massachusetts native, Davis had earned a noteworthy reputation as an ensign in the Civil War during one of the Union sieges on the Confederacy's Fort Fisher. That day, the young naval officer made it farther than any other person on the Union force, charging on foot through a hail of bullets as he took down a sharpshooter.

In commanding the *Juniata*, he now stood at the helm of one of the warships that decades earlier had been part of that squadron at Fort Fisher. The *Juniata* was named for a river in central Pennsylvania and had proven its value while providing military support to Union general William Sherman, as well as capturing ships supplying the Confederacy.

By the time they received their instructions regarding Midway, the crew of *Juniata* had spent three and a half months in Honolulu. The mood there had been lighthearted. A group of *Juniata* cadets spent most of their time gossiping about Hawaiian girls. Sailors from the sloop also played baseball, an increasingly popular sport in Hawaii, against the crew of yet another U.S. vessel, the *Vandalia*. During this period, one of the seasoned gunners on the *Vandalia* was William Halford, the lone survivor of the *Saginaw*'s lifeboat. During lulls in the baseball game, Halford, now forty, could tell the *Juniata* crew all about Midway, the unassuming atoll that seemed to sabotage every effort to fulfill the *Saginaw*'s mission, and the deprivations they suffered upon wrecking after leaving Midway.

Though they fell short in baseball, losing 18–1, the *Juniata* crew fared much better in a regatta, taking first prize in a race of small boats. As the crew competed for bragging rights against other American sailors in Hawaii, their officers were invited to formal events. Commander Davis was among the notable attendees at the funeral of a member of the Hawaiian royal family, along with British commissioner James Wodehouse. Deputy Marshal Pahia, who had been sent onto the *Minstrel* by Wodehouse during the near-mutiny, guarded the large funeral procession on horseback. Commander Davis and his officers were also presented for a private audience with the king. In turn, the American visitors hosted Honolulu society for a reception and dance on board the *Juniata*, and the king welcomed the officers for a palace dinner before their departure.

Now the social calendar was cleared and the naval steamer was heading for Midway.

Juniata was an ideal vessel for managing treacherous passages. In fact, part of its purpose was to make travel safer for other ships, as demonstrated when Commander Davis had sailed *Juniata* around the world taking soundings for depth. The screw sloop (that is, powered by a screw propeller) also had a record assisting in rescues. A decade prior, the *Juniata* sailed to aid *Tigress* in seeking survivors of the wreck of *Polaris* in the Arctic with orders to search "until the last moment of the navigable season." On a voyage a few years later, the *Juniata*'s crew brought a German schooner to safety.

The *Juniata* had left Honolulu three weeks to the day after the *Wandering Minstrel* had sailed from there, and the warship possessed a particular advantage over a schooner in what one of the sailors aboard called their "good iron horse"—the steam power provided by two engines. One of *Juniata*'s sailors prepared for their planned stop at Midway, charged with documenting for navy records "any changes which might have occurred around [Midway's] coral and uninviting shores." While not glamorous, the assignment would give the sailors a unique experience as the first official United States naval visit to the atoll in decades—and though they could not know it, they were also poised to be heroes for saving a family of castaways and their crew.

After twenty-five days, they were closing in on an estimated forty to fifty miles from Midway. The weather took a turn, becoming squally, with cyclone-like conditions. They were determined to continue, but their sails were knocked out one by one. Reality dawned—not only might they never make it to Midway, but the sloop might end up lost altogether. Nor did they have any way to communicate while out at sea.

Soon, the United States Navy, with no word on the warship's status, reported the *Juniata* as missing—apparently another casualty of trying to tame little Midway Atoll.

Meanwhile, Commander Davis and his officers recognized that the rough conditions would amplify the "dangers surrounding the island." Davis decided they had to give up on Midway, turn and seek the trade winds, prevailing east-to-west winds that ships often relied upon when encountering storms.

Even after leaving behind the worst weather, high-force gales from the north still hampered them. Davis had to consider when and where they could restock the coal for their engines, but two months into their voyage the only land they had sighted were two volcanic islands, one of which was in the midst of an eruption. To make matters worse, weevils ruined their flour and rice, reducing rations largely to "salt junk," or dried beef. To pass the time, the crew played cards, wrote diary entries, and took photographs with cameras stowed on board. Below deck, studying their charts in somber conversations that felt like "a deliberate council of war," the commander and his officers determined they might have no choice but to head for the Bonin Islands, ordering the sailors to prepare axes for chopping wood there, "where we might [remain] indefinitely." Within sixty miles of Bonin, *Juniata*'s officers found a favorable wind that allowed them to redirect the sloop toward Yokohama, which they finally reached after seventy-nine days at sea.

Upon anchoring, a small boat from USS *Monocacy* carried Lieutenant Albion Wedhams, who rushed to board the *Juniata*. Wedhams gave them "the startling news" that they were considered lost.

If Davis had succeeded in finding a way to reach Midway after the unexpected change in course, factoring in the delays caused by weather, the *Juniata* would have arrived within weeks of the *Minstrel*'s wreck. But even a formidable sloop of war with a pair of steam engines, a commander celebrated for heroism by Abraham Lincoln, and sailors who had recently shown their prowess in a regatta, may never have made it through the conditions around Midway.

"Our escape," wrote one of *Juniata*'s officers about their ill-fated odyssey, "was a genuine god-send."

THE WALKERS WERE watching the horizon for any sign of a sail. Their emotional experience evoked the words written by one of the *Saginaw*'s stranded lookouts: "[I] searched with anxious eyes through my rescued opera glasses the shipless horizon; sometimes with such a strain of nerves and hope that phantom vessels plague my vision. The loneliness and solitude of the vast expanse of water surrounding us is beyond expression. . . . One might easily fancy that beyond the line of the horizon there exists only infinite space." The *Wandering Minstrel* castaways also kept fires going whenever possible. However unlikely it was to be spotted, there was little else that they could try, though over the years castaways developed ambitious ideas. After one 1842 shipwreck, castaways on another Pacific island caught albatrosses and tied wooden shards inscribed with messages to the birds, hoping they would fly far enough to reach rescuers (which they apparently did not). There was no evidence the *Minstrel* castaways tried this. But neither could the Walkers sit idly and wait for a miracle.

A plan came into view: They would improvise a vessel using parts of the damaged lifeboats, the pieces of the *Minstrel* wreckage gradually washing up ashore, and the small supply of extraneous material deposited on the island.

All the while, the albatrosses began to fly to other locations for the season, which meant the castaways needed to locate the harder-to-spot soft bird they called mutton or "puppy birds," the wedge-tailed shearwaters, to use for food. The bird was difficult to find because it burrowed under the sand. Jessie, the dog, became their primary hunter, tracking and rooting out the birds from underground. "If there was no bird in the hole," Elizabeth observed, "Jessie would sniff and go on to the next." Jessie, saved from the sinking ship at the last moment, was now saving their lives.

The castaways could be divided into two basic categories, each of which boasted a strength that was also a handicap. In the first group were those who adopted John Cameron's "every man for himself" attitude, which gave each a freedom from feeling responsible for anyone else's welfare, but also meant they could not count on someone looking out for them. One of these castaways began to hoard food and to threaten those who questioned him. In the second group, the Walkers and a faction of the crew relied on each other implicitly, but that meant every decision was unwieldy, weighed carefully for its collective impact. Jessie stayed at the Walkers' side, in contrast to some wrecks in which ships' dogs were reported to have become feral.

The escape plan, meanwhile, called for a few sailors to take the best vessel they could construct to sea, in hopes that they could reach help to send to the rest. There was precedent for this approach. In 1822, two British whaling ships sailing on a coordinated voyage, the *Pearl* and the *Hermes*, wrecked on then-unnamed reefs and castaways used the wreckage to build a ship they optimistically christened *Deliverance*, which successfully sailed to Honolulu. A more recent example came from the coal bark *Henry James*. After rowing to Palmyra Island, five sailors launched one of the lifeboats, reaching Samoa and sending help for the rest.

However, there were differences between those examples and the *Minstrel*. The *Henry James* castaways had access to relatively extensive supplies with which to fill a lifeboat, and the wreckages of the *Pearl* and the *Hermes* provided a wealth of materials. The *Minstrel* castaways had to rely instead on experience and ingenuity. George Hanker, the longest-serving sailor among them, helped with the construction of the boat, but began to withdraw into solitude without explanation.

As captain, Frederick could have dictated that the Walkers would be the ones to take the expedition themselves once the boat was completed, but that would leave no room for anyone else. Moreover, it would strand the others on the island without a clear leader at a time when emotions and alliances were fraught. It was a dilemma with no easy solution, but the Walkers believed "family" applied to everyone under their protection. The Walkers would stay put, and the finest sailors would be selected from the rest of the castaways for the all-important voyage.

The knowledge gained from Hans Jorgensen continued to be especially useful as their plans took shape. For example, Hans knew Sand Island was not suitable as a launching point because of its uneven, ragged shape, but that Midway's smaller land formation, Green Island, while no more habitable, presented a safer entrance into open waters. Frederick also found Green Island a "pretty island to roam about" where they could go "prospecting for food," namely fish and eggs at times when Sand Island ran low, although every step through the rough brush on Green Island could slice up arms and legs.

Frank Lord, the cook whose life was saved after capsizing in the lifeboat, became impatient with all of this. He could be impulsive and volatile, having been in trouble with the law in Honolulu for drawing a revolver during an argument. Frank and his clique that included four like-minded crew members quietly hatched a scheme. They swiped provisions that had been rationed for the whole group, and secretly started salting fish and bird flesh to remain palatable longer. They would make their own escape.

Chapter 7

SHIPWRECK OF FAITH

March 15, 1888

THE MORNING OF MARCH 15, 1888, someone noticed that several of their shipmates, Frank Lord included, were gone. Nightmare scenarios presented themselves: the castaways, if they had slept close to shore, could have been swept away by a rogue wave; or one of the wells they had been digging may have caused a sinkhole, which swallowed them into the ground. The remaining castaways launched a frantic search.

"Soon," wrote first mate John Cameron, "we observed their boat rounding the western point of [Green] Island, under full sail with a fresh breeze blowing; it headed for the passage and the open sea." Frederick spotted it too. He rushed to prepare another lifeboat to intercept the stolen vessel. But the remaining boat had not been reinforced and the vessel was already out of reach.

The castaways quickly pieced together what had happened. Under the cover of darkness, Frank's group had made their way across the lagoon to Green Island unseen. They took the vessel the whole group had been repairing. It was the "best boat we had," Frederick lamented, on which they had pinned their hopes. Frank and his allies had also brought the fish and bird flesh they had swiped and salted.

Amid the rush of emotions, there was a brief glimmer of optimism that the absconding crew members might make it to safety and, despite their deception, send help. But unlike the successful examples of the *Henry James* lifeboat that reached help from the wreck at Palmyra Island and the *Deliverance* sent from the double wreck of the *Pearl* and *Hermes*, there were many unhappy precedents. In the wreck of USS *Saginaw* on Ocean Island, only William Halford, now of *Vandalia*, survived the journey to Kauai, Hawaii, and just barely.

On Sand Island, the plan hatched by Frank and his cohorts had been flimsy. They had no compass. The small vessel was not yet deemed seaworthy, and was helmed by a cook who earned that position because of his poor showing as a sailor.

MEANWHILE, IN WASHINGTON, D.C., State Department officials heard about USS *Juniata*'s unexpected arrival in Yokohama after failing to reach Midway. Assistant Secretary of State George Rives, thirty-eight, had been appointed to his position a year earlier, and like his superior, Secretary of State Bayard, came from a political family. Rives could request that the U.S. Navy send another vessel from the Pacific squadron to complete the aborted surveying operation at Midway. But Rives argued that Midway "was of no value," rebutting Secretary Bayard's open-minded perspective on the atoll. Ultimately, government officials leaned toward Rives's position and decided not to send another vessel. The *Juniata*'s experience, on top of everything else, counseled that an excursion to Midway was not worth the calculated risk. How could

Midway be utilized for practical purposes, if it could not even be reached?

With shipping so crucial to commerce and government around the world, reporters chronicled vessels' arrivals and departures. But neither the press nor the public expected quick turnarounds: the *Minstrel* could be gone for several months without prompting any concern, so nobody was looking for the Walkers. In the meantime, the Yokohama-based *Gazette* incorrectly reported that *Juniata* had reached Midway, and this error was reprinted in the San Francisco and Honolulu press, the two places where eventual searches for the *Minstrel* would have been most likely to begin. For those who mistakenly believed *Juniata* had anchored at Midway, they assumed that nothing interesting was to be found there. Thus Midway could be the very first location crossed off a list of potential places to look for the *Minstrel* crew. Still, the error would please the State Department, which might even have facilitated it, as it helped establish a public record, however misleading, that the United States had not abandoned possession of Midway.

Unbeknownst to the castaways, there was another vessel at sea that represented the possibility of rescue. A full-rigged iron ship called *Halewood*, owned by the Leyland brothers of Liverpool, sailed from Newcastle, New South Wales, toward San Francisco. Though lacking the resources of U.S. Navy ships, the *Halewood*, built three years earlier, was in good condition. Captain Henry Stap made no secret of seeking glory at the helm.

The son of a lieutenant in the British Royal Navy, Stap, fifty-nine, was a seasoned merchant mariner who liked to tell stories of his career. Few better examples of his highs and lows existed than his first time in command, on a vessel called *Mystery* in the early 1860s. In 1861, Captain Stap sailed *Mystery* from London to New Zealand in what was then considered record speed, taking seventy-five days. But just a few years later, *Mystery* foundered in a powerful storm, with Stap almost drowning when he was the last one to be pulled from the sinking ship by a passing vessel. Later, he took pride in his five years commanding what was said to be the largest three-mast ship on record at the time, the *Great Britain*—though that also ended up wrecking. His experiences gave him a philosophical and scrappy air. "His clear eye is that of a younger man," observed a reporter who met him in Washington Territory, "and his step is firm as a soldier's."

Halewood's latest passage through the Pacific, toward San Francisco, progressed largely as planned until around the time Stap logged latitude 54 south, longitude 168 east, with the closest land being the uninhabited, mountainous Campbell Island, where a whaling boat vanished years before. *Halewood* was caught in gusts that pushed the vessel north. Buffeted by the conditions, Stap studied his charts to find safe harbor, with all other islands in range inaccessible or unsafe due to terrain or political conflicts.

Crossing the equator, Captain Stap correctly calculated that they came closer to Midway Atoll. He could not know, but upon reaching Midway, he would have collected another of the "sea stories" he cherished, as there waited a family of castaways and their shipmates who nobody yet knew were missing. The reef would be treacherous, but after being known for commanding the massive *Great Britain*, if anyone could reach Midway, or sink to the bottom of the ocean trying, it was Captain Henry Stap.

But a hurricane, as Stap would describe it, "sprang up" as though out of nowhere, and like the *Juniata*, the *Halewood* lost several sails. Stap had to change course again, finding another way through and out of the storm, bypassing Midway altogether. By the time *Halewood* reached San Francisco, Stap had set a record on the other end of the spectrum from his previous record on the *Mystery*—the *Halewood*'s 103 days at sea was now reported to be the slowest passage on record for a ship sailing from New South Wales to San Francisco. Perhaps motivated to distract from the indignity, within weeks Captain Stap publicly announced that *Halewood* would challenge another ship in a race to Cork, Ireland.

That fact attracted a modest amount of interest from the press. In the process, Stap spoke to a reporter about Midway. He said he "would not advise anyone to try" going to Midway until seasonal conditions changed, warning against the voyage for at least three or four months. His comments would dissuade any mariners who had a stop at Midway in mind—another blow to the odds that anyone would find the Walkers.

NONE OF THIS could be known on Midway, where they tried to keep the bonfires blazing for any passing ships. But those fires were being doused from above. The same night that Frank Lord and his conspirators absconded with the reconditioned lifeboat, a heavy gale swept in. The castaways on Sand Island watched from the shore as the sea turned fierce. Grief subsumed the feelings of anger toward Frank and the other rogue sailors. No question about it: their former shipmates would drown.

The remaining castaways could benefit from any distraction from their circumstances, though distractions were hard to come by. By one account, the only book that had been brought from the *Wandering Minstrel* onto Sand Island was *Uncle Tom's Cabin*, the widely read Harriet Beecher Stowe novel about life among enslaved people in antebellum United States. Decades removed from the Civil War, the story tended to be read as a universal text on morals and personal freedom, and the book achieved wide readership in Great Britain. "No other book ever printed in any age, country, or tongue," wrote a journalist at the time the book comprised the castaways' entire library, "with the sole exception of the Bible, has ever had a circulation or an influence at all comparable with 'Uncle Tom's Cabin.' Hundreds of millions of people, in every clime under the sun, have read it."

Though the castaways had no Bible, Elizabeth knew scripture well, and others looked to her to share her strong faith in God. She could relay the Bible's verses on shipwrecks, including Saint Paul's accounts of his suffering: "Thrice was I beaten by rods, once was I stoned, thrice I suffered shipwreck . . . in journeyings often, in perils of waters, in perils of robbers, in perils by my own country-men, in perils by the heathen, in perils in the city, in perils in the wilderness, in perils in the sea, in perils among false brethren." Hardships at sea and shipwrecks, which bore repetition in Paul's statement ("shipwreck . . . perils of water . . . perils in the sea"), evoked other biblical episodes Elizabeth could impart and interpret for the castaways.

One such storied voyage taken by Saint Paul, which ended with yet another shipwreck that occurred after the other three, also included Paul's friend and fellow apostle Luke. As described in the book of Acts, Paul was present on the ship as a prisoner and warned the captain and crew against sailing. "Sirs, I perceive that this voyage will be with hurt and much damage, not only of the lading and the ship, but also of our lives."

The captain ignored Paul's admonition and the ship later sailed through severe tempests, causing the vessel to run aground and break apart. The guards decided to murder the prisoners so they could not escape by swimming off, but their commander stopped this barbaric plan. In a vision, Paul had received a message from God that they would all survive. They ended up washing ashore on a "certain island." The islanders first believed there were signs that Paul was a murderer, then that he was a divine leader.

The narrative contained several elements that could be recognized in the *Minstrel* castaways' circumstances. The story captured a belief that perils at sea doubled as tests of faith. "Let it be remembered," preached a popular English minister regarding this biblical episode, "that the salvation of the soul is of infinitely greater importance than the salvation of the body." Elsewhere, in a passage attributed to Paul, was the notion that people could "make shipwreck of faith," or let faith falter in adversity. As Frederick noted often happened with sailors in trouble, "any religious training they had when children was apparently obliterated by their surroundings."

The stolen boat—and the presumed drowning of their six shipmates on board—provided a sobering turning point for the Walkers on Midway, who tried to count small blessings. Three months into being stranded, they gave thanks for the discovery of additional debris from the *Minstrel*. Elizabeth recalled that "there came ashore three mattresses, some blankets, five plates and five knives and forks." She considered how the latter number fortuitously matched the number in their family, though they did not have much use for silverware. Walking along the beach, she also found a "soup tureen" in the sand, and an iron kettle, which apparently came from another wreck. More helpful for the nourishment and morale of the castaways, a container of potatoes washed up.

Another rediscovered object was a tin box that belonged to Elizabeth, which she had assumed had been lost on the ocean floor. The box contained a silver tea set that she had adored as a child. When objects had first started to show up on Sand Island with the tides, she had "longed" that by some small miracle her tea set could be found, and now she cradled it in her hands. Clinging to a sign of home, Elizabeth could share her family treasure so full of memories with her sons.

Elizabeth became the cornerstone of life on the island. With Frank Lord gone, she used her cooking skills to turn the odd scraps into delicious meals for her family and the other castaways. When a cask of rice washed up on land, but was too fermented to eat, Elizabeth mashed it and fried it in "gooney oil" made from birds. The result was divided up and relished. She also provided emotional sustenance: it fell to Elizabeth to urge patience and positivity on men teetering on the edge of despair.

They were grateful for Hans's continued help as they regrouped after Frank and his clique vanished with their vessel-in-progress. Hans was a carpenter, which could be key to fitting up another seaworthy vessel with their depleted supplies. As Hans became integral to the castaways' activities, some thought of him as serving the second mate role now that George Hanker had retreated into solitude.

July 4 brought memories of how the Walkers usually marked the holiday in more carefree times at sea. They would decorate the ship with colorful cloth and sing American anthems as well as Irish ones from Frederick's homeland, for good measure. "As we had no copy of the Declaration of Independence to read," Frederick recalled, "we whistled the Star Spangled Banner, shouted E Pluribus Unum, Erin go bragh, until eleven o'clock a.m.; we then drank to the health of the President of the United States, all hands giving a hearty cheer."

Normalcy and optimism, as well as ritual and routines, became increasingly difficult to maintain on the island. In a kind of mental torture, sails occasionally appeared faintly on a distant horizon, brief sightings of ships like the *Juniata* and the *Halewood*. The castaways would do everything they could to signal the vessels, to no avail, leaving them in a cycle of what Elizabeth felt as "bitter disappointment." The amount of time that they could realistically expect to survive dwindled. Illnesses set in, which meant fewer people able to hunt for and gather resources, a number already diminished by the absconders. Inevitably, order threatened to unravel on the deserted island. While the temporary shelter village was fortified with debris, the Walkers' use of the sturdiest hut also provoked resentment.

Meanwhile, the bitterness that had been directed at the crewmates who stole the boat was increasingly channeled toward Hans. As birds migrated away for the season, the sailors coveted Hans's preexisting supply of eggs. Hans's unusual hardiness stoked further antipathy. Despite his longer state of deprivation, he remained energetic and muscular, and seemed immune to illnesses plaguing the newer arrivals.

Frederick had his revolvers from the *Minstrel* and would be prepared to use force if necessary to protect people on the island, as well as their personal possessions, including Hans and his hard-earned cache of food.

The castaways had reasons for hostility toward Hans that went beyond their own hunger and desperation.

He began to behave in peculiar ways, sometimes with rage-filled outbursts, which fomented distrust in his new companions. Perhaps the guardedness Elizabeth had developed for strangers over her years of sailing had to be applied with more vigilance to Hans, as the man they had first looked to for salvation revealed his pent-up anger. Slowly, details slipped out that indicated Hans was very different from the person they thought he was. He hadn't just randomly washed ashore onto Sand Island. He had been marooned there intentionally by another ship whose sailors knew the truth: Hans was a murderer.

Chapter 8

HANS'S STORY
PART I

1886

THE PATH THAT LED HANS JORGENSEN to the Walkers had begun when the Danish sailor was employed by a German cargo vessel called *Apollo*, which in 1886 transported sugar between Honolulu and San Francisco, a small cog in the large wheel of Hawaii's commerce with the United States. Hans deserted the ship in Honolulu, which led to his arrest through the influence of the Imperial German Consul to Hawaii, Henry F. Glade. Glade, forty-two, had come to Honolulu from Germany (where he was called Heinrich) five years earlier as a partner in a mercantile firm, and after proving himself to the community as "a man of sterling integrity and uprightness," became Germany's consular agent to the kingdom. He and his wife, Clara, enlivened the Honolulu social scene, throwing galas at their home with elaborate decor, music by the Royal Hawaiian band, and guests that included the king of Hawaii.

Proving he could oversee orderly shipping operations helped Glade's career in Honolulu, which included ambitions to establish a telegraph cable connecting Hawaii and North America, a potentially massive boon to trade. For Glade and his partners in the telegraph enterprise, there was "no such word as fail." Now he had to make an example of a deserter such as Hans. Of course, Hans disputed the accusations against him. He said that *Apollo*'s captain had withheld rations in order to starve the sailors until Hans led a mutiny.

Glade was unconvinced. But Hans managed to disentangle himself from Glade and Hawaiian officials long enough to join the crew of a schooner, *General Siegel*, which prepared to launch a shark-fishing expedition. It was no surprise that the owner of the *Siegel* was looking to reverse his fortunes through sharking, like the *Wandering Minstrel*, as the ship had become a money pit. The schooner had gone through multiple captains in a short span of time and always seemed to require costly repairs. In fact, during a recent voyage, a representative of the ship contemplated selling the *Siegel* while sailing, never to return to Hawaii.

Inconsistent oversight and hesitancy by ownership could well have contributed to slipshod hiring practices, thus allowing Hans Jorgensen to join the *Siegel*'s eight-person crew. At the same time, the captain, who went by the name Frank Sanders but whose surname was really Asberdine, worked under added pressure, knowing that the ship could still be sold. Sanders had been an officer on the vessel before his promotion to captain and may well have felt greater attachment to the ship than the owner did. Joining him was a respected older Honolulu sailor named Peter Larkin. Peter was dealing with his own financial setbacks, with authorities forcing him to sell his sailboat, *Sarah*, because of a $173.52 debt. Only two months after the involuntary auction, Peter could take bittersweet satisfaction in reading in the Hawaiian *Daily Bulletin* that the new owner ran *Sarah* aground, leaving behind a total wreck.

On September 1, 1886, the *Siegel* sailed for the French Frigate Shoals, a logical destination considering the frequent shark sightings around land formations such as Shark Island. The crew members reported a successful haul. Almost a month into the trip, they had reached the Midway Atoll, approximately sixteen months before the *Wandering Minstrel* wrecked there, with plans for an extended stopover to fish and collect eggs before continuing their passage.

With the schooner anchored, the eight sailors camped out on Sand Island. The *Siegel*'s sailors at first welcomed a change of scenery from the confines of their vessel. As the men got to know each other better, Hans revealed the grudge he nurtured against Henry Glade, the German consul in Honolulu, because of the clash over his desertion from *Apollo*. More than once he said that he would kill Glade, a threat even bolder considering how well connected the wealthy Glades were in Honolulu. Captain Sanders reprimanded Hans for the inflammatory statements.

During the night, hours before they planned to sail away from Midway, gale-force winds enveloped the island. The anchored ship was thrown against the reef. Upon awakening at sunrise, the men could see that the *Siegel* had begun to break apart and sink. Sanders and his crew realized they were stranded.

By this point, they had already familiarized themselves with Sand Island. They took turns sheltering in the hut left behind by the American coal venture. The castaways concluded—as the Walkers would later—that their only sustenance would come from catching fish and seabirds, and collecting eggs.

Sanders assessed the situation and announced that they had to wait on Sand Island "until a rescue party should come from Honolulu to take us back."

Meanwhile, approximately ten days into being stranded, Hans's *Siegel* shipmate, Peter Larkin, decided to fish with dynamite. The method, said to have originated with miners, stirred controversy because of damage to coral reefs and marine life. "For one fish caught for food in this way," observed a chronicler of fisheries at the time, "a hundred were wantonly destroyed."

During Peter's attempt, the dynamite exploded and blew off his right hand. The rest of the castaways tended to his injury, worked to stop the bleeding, and tried to make him as comfortable as possible.

The gruesome accident was a reminder of the compounding effects of desperation and exhaustion among castaways. It remains unclear whether the *Siegel* crew brought the dynamite ashore with them, but more likely it had been left by the aborted American coaling expedition years earlier. Explosives including dynamite became less stable over time, making an already unsafe practice riskier.

Following days of horrific pain, Peter died. The death put the already bone-weary and emotionally drained castaways further on edge.

The spare scow from the *Saginaw* was on the island but was beyond repair. Edvart Olsen, one of General Siegel's seven remaining castaways, found another lifeboat that drifted ashore. It was an astounding development, all the more so because it linked to yet another shipwreck. The unmanned lifeboat was marked with the name *Dunnottar Castle*, a wreck from four months earlier that proved rather painless to all involved, with the exception of the Hawaiian government.

THE CALIFORNIA-BOUND *DUNNOTTAR CASTLE*, carrying a cargo of coal, had wrecked on July 15, 1886, after a chronometer malfunction led to a collision with the dangerous reefs of Ocean Island that long before claimed the *Saginaw*. Seven sailors rowed one of the lifeboats to sea and reached Hawaii after almost two months, while the rest of the castaways found safety in half that time when they were plucked from Ocean Island by a passing vessel. Meanwhile, one of the other lifeboats from *Dunnottar*—this one with no one inside—floated with the tide, ultimately to end up at Midway.

The *Dunnottar Castle* wreck demonstrated the complex dynamics of rescue missions, which would be relevant to shipwrecks still to come, including *General Siegel* and *Wandering Minstrel*. When the contingent of castaways from the *Dunnottar* arrived in Hawaii in their lifeboat, they pleaded for the rescue of the remaining castaways on Ocean Island. The Hawaiian government worried that another nation's vessel could reach the *Dunnottar* castaways first, in the process claiming possession of the island. So Hawaiian officials coordinated with James Wodehouse, the British commissioner to Hawaii, to charter a steam-powered vessel called *Waialeale* from the Inter Island Steamship Company to sail for Ocean Island. The gleaming steamer, 130 feet long and 35 feet wide, had launched for the first time only weeks earlier, with crowds cheering at ports along the way. Circumstances seemed almost to mock the would-be rescuers' urgency. Instead of leaving in a hurry, sailors first had to carry off five hundred bags of bone meal, or dissolved animal remains, cargo that had been loaded the day before it was chartered.

King Kalākaua came to the wharf to see the *Waialeale* off. He was never shy about making public appearances, and he was credited as the first monarch to tour the globe. "He is a remarkably handsome man," a reporter in New York had observed at that time, "more than six feet in height. His carriage is erect, his beard, mustache and eyes are black, and his expression is amiable." He had assigned one of his special commissioners, James H. Boyd, twenty-eight, son of a British father and Hawaiian mother, to accompany the *Waialeale*'s mission, and instructed him to find time, in between rescuing castaways, to claim Ocean Island in the name of the Kingdom of Hawaii. The Hawaiian delegation included an artist, Joseph D. Strong, to sketch scenes of the rescue. Strong, as it turned out, had reason to be distracted from his assignment, as his marriage to Robert Louis Stevenson's stepdaughter faltered due to Strong's infidelity. Meanwhile, the Inter Island company contributed to *Waialeale* four live sheep, presumably to be fed to the castaways on the island.

But neither Wodehouse nor anyone else involved had guessed that the remaining *Dunnottar* castaways they set out to save were long gone, having been rescued by a passing ship (*Birnam Wood*, like *Dunnottar Castle*, was named after a locale in Scotland), as relayed by a series of notes on the island. The only castaways left were three dogs that had been aboard *Dunnottar*. Befuddled, Special Commissioner Boyd and his team of Hawaiian representatives busied themselves officially claiming possession of the deserted island and tending to the dogs and sheep. In late September 1886, they also assembled a shed with tanks to catch rainwater for future castaways, and devised an unusual plan to bury meat for hypothetical castaways.

TO THE *GENERAL SIEGEL* castaways on Sand Island, it seemed to be a miracle that a boat marked with *Dunnottar Castle*'s name from a shipwreck fifty nautical miles away had traveled to the site of another. The lifeboat was in poor shape, and likely resembled another *Dunnottar* lifeboat, the one that managed to reach Hawaii with sails "rather flimsy for ocean blasts." But without any sign of rescue, the *Siegel*'s Captain Sanders wavered on his determination to hunker down and wait. They needed to consider other ideas to get off the atoll, including the risky proposition of using a vessel never intended for a long ocean voyage.

Sanders was right to rethink his earlier position. A search mission to find them was anything but a given, even after enough time had passed for the "mystery" of *General Siegel*'s whereabouts to be discussed back in Hawaii. "Is there to be no search for them?" asked a reporter in a column in the Honolulu *Evening Bulletin* headlined "Where Is the General Siegel?" "Are they to be simply numbered with the missing? Surely six or seven men's lives are worth chartering a schooner and sending in quest of the missing men." In addition to challenging the Hawaiian government, the columnist questioned why the owners of the *Siegel* did not send one of their other vessels to search the Pacific islands.

A lost or missing ship often triggered a political tug-of-war. Questions of which government entity was responsible, if any, to search for vessels took into account the variety of nationalities of an owner (or group of owners or company), the captain, and the officers and crew members, who often hailed from all over the world. Captain Sanders himself represented a medley of heritages, being part Russian and part Finnish, and to further blur the facts, even those who knew his given name were not sure how to spell it, using variations of Asberdine. Searches required large sums of money, and in at least one case in recent memory (the U.S. Navy's *Saginaw*) a ship wrecked while looking for lost ships. The Hawaiian government already faced scrutiny for recently spending money to upgrade a naval ship, christened *Kaimiloa*, while the kingdom's roads and infrastructure remained underfunded, and critics now demanded that this naval vessel be sent to search for the *Siegel*.

None of this was known by the *General Siegel* castaways on Sand Island, who found yet another small boat, a sampan, abandoned by a Japanese schooner called *Kaulilua*, which had stopped at some point in the past to fish near Midway. In addition, Sanders and his crew had the lifeboat they had rowed from the *Siegel* before the schooner ran aground on the reef. These small boats, at least, allowed them to cross the lagoon between Midway Atoll's two islands, Sand Island and Green Island, a passage the *Minstrel* castaways would routinely make more than a year later.

None of their small vessels was safe to sail any significant distance, but Sanders settled on trying to repair and reinforce the *Dunnottar* lifeboat the best they could.

As reality set in that they would be there for a long time, one of the sailors, Will Brown, carved a message on the wall inside the wooden hut, in part to memorialize Peter Larkin's death. If Brown already worried about what Hans might do, he did not leave any hints in his carved message.

ARRIVED SEPTEMBER 1886

W. BROWN

HAW.[HAWAIIAN] SCH. GENERAL SIEGEL

HONOLULU

1886–1887

LOST ON THIS ISLAND THE NIGHT BETWEEN

16–17 NOVEMBER 7 MEN IN 1886 SHORE.

1 MAN DIED AND BURIED IN SHORE 7TH OCTOBER '86.

As time passed, Hans became increasingly agitated. Apparently, he had convinced himself that if they escaped and reached Hawaii, his crewmates would turn him in for his threatening comments toward Glade. A darker interpretation would be that the threats were not mere threats at all, that Hans actually planned to murder the consul, and that he worried that his crewmates could interfere with the assassination. Certain comments made later by Hans suggested the possibility that while Peter Larkin was convalescing from his injuries from the dynamite, Hans had poisoned him.

William Timpe, nineteen, the youngest of the *Siegel* castaways, had heard those threats by Hans against Imperial Consul Glade, and William, like Glade, happened to be German. One day, the young man went into the hut on Sand Island to collect stored eggs to prepare for a meal. Hans walked in quietly behind him and threw a rope around William's neck. Hans tightened the rope, stopping him from making any sound. In the "infinite space" represented by the tiny island, Hans's capacity for violence escalated, with a growing sense that no consequences existed for his actions. The young man, in search of breath, twisted and thrashed. In the tussle, the rope came loose. William fell to the ground, gasping for air.

Hans backed away from him. "Don't say anything about it," he warned William, referring, presumably, to both the threat against the German consul and the attempted strangling. "I don't want to do you any harm."

Chapter 9

HANS'S STORY

PART II

1887

WHILE URGENCY MOUNTED SURROUNDING HANS'S behavior among the *General Siegel* castaways, back in Honolulu in the spring of 1887, the owners of the missing schooner finally commissioned a vessel, the three-masted *Ke Au Hou*, under Captain Brownell, to conduct a search for the *Siegel*. Debates remained heated when the Hawaiian government declined to fund their own search, and the *Siegel* owners pointedly did not ask the kingdom to subsidize the *Ke Au Hou* search party.

On Sand Island, Sanders and the other castaways, while trying to keep Hans's troublesome side at bay, were repairing the *Dunnottar Castle* lifeboat that had floated onto their beach. When they finished their alterations, they found to their dismay that they could not all fit into the small vessel. Sanders decided they would also prepare the *Kaulilua* sampan, so they could launch both boats simultaneously in order to accommodate all seven of them. After spending time modifying the sampan, they watched with heavy hearts as it capsized during a test.

"We then put a ballast in her," according to Edvart Olsen, who worked on the sampan, "but she failed to keep afloat." They turned their attention to a mast of the *Siegel* that they had recovered from the wreck. "We took the schooner's mast, cut it into short lengths and fitted them to the side of the boat to act as [outriggers]," or beams intended to stabilize the sampan.

While Edvart, William Timpe, and two others worked on the boats, Captain Sanders crossed the lagoon to Green Island with the other two crew members—Hans and Brown—using the smallest vessel, the *Seigel* lifeboat. It is unclear if William had concealed Hans's assault from the others for fear of reprisal, but if he had, Sanders may not have been as cautious around Hans as he otherwise would have been. In addition to collecting timber for their ongoing repairs while on Green Island, Sanders wanted to bolster their resources, especially their food supply, before they attempted their escape from Midway.

Sanders, Brown, and Hans went over to the smaller island on Wednesday, June 15. It is impossible to know exactly what happened next. Hans had a Winchester rifle with him. He may have waited until the men split up on their quest to gather anything useful. If Hans plotted out whom to target first in order to retain the element of surprise, he chose the more formidable of the two, Captain Sanders, who was armed. Hans shot him dead. Brown, who was unarmed, may not have become suspicious when he heard gunfire, since they had come prepared to hunt birds to stock up for their departure. For this reason, Brown was not on alert when he returned to their rendezvous point. Hans killed him, too. *The Wrecker* includes a guileless character named Brown who also becomes a murder victim, futilely begging for his life when he "poured forth out of the darkness an endless, garrulous appeal for mercy."

Later that afternoon, Hans returned alone to Sand Island. He nonchalantly asked the others for the time, and declared that it was getting too late in the day to return to Green Island, where he said the captain and Brown decided to stay in an encampment to continue their preparations. In the morning, Hans returned alone to Green Island, this time staying for ten days. This gave him time to decide how to dispose of the bodies—either to weigh them down to sink in the ocean, where they would likely be consumed by sharks, or to bury them. He also had time to clear away other evidence of bloodshed.

Back on the larger island, Edvart and the others surveyed Green Island from across the lagoon that divided the two land formations. They could see the lifeboat Hans had used to go back to Green Island but could not see Hans or anyone else. Edvart thought he should investigate, but because they were repairing the other two boats, they could not safely get to the other island across the shark-filled lagoon.

When they completed their latest repairs on the sampan, they tested it on the water. Hans apparently watched them and realized the others could easily and quickly make it to Green Island.

Suspicions, meanwhile, increased with every passing day that failed to bring any sign of the captain and Brown. Edvart searched Hans's hammock and found a loaded revolver and shotgun, and something else equally unnerving: Hans's notebook. An entry read: "Six men left for Marshall Islands at the beginning of June." They were now in late June, and nobody had left the island. The gambit was clear. Hans planned to murder them all and take one of the repaired vessels himself to Honolulu, with the faux journal entry left behind as misdirection in case anyone came to the island in search of them. Since the rest of the castaways would by that point never have shown up in the Marshall Islands, it would be assumed they had drowned on the way.

All the while, the *Ke Au Hou* had been stopping at a series of Pacific islands, anchoring and sending boats ashore when possible to look for the *General Siegel*. At Ocean Island, they documented what remained of the earlier wreck of the *Dunnottar Castle*. "Nothing was left to be seen of her except a single spar sticking up out of the water, and a mutilated seaman's chest on the beach." They anchored at Laysan Island, filled with birds and ducks, and also went ashore at the remote, low-lying Lisianski Island, where they found pieces of what they deemed "ancient wrecks" but nothing of the *Siegel*. They came upon a warbler that they brought back onto the ship as a kind of pet. Remarkably, they sailed right past Midway Atoll, where the two land formations of Sand Island and Green Island had become the scene of life-and-death conflict. But with no sign of the *Siegel* visible from their position, they did not risk navigating the reefs, and ultimately sailed back to Honolulu with no clues to the *Siegel*'s whereabouts.

Back on Sand Island, around nine o'clock at night in the pitch-darkness, Hans approached the hut. He appeared ragged, missing part of the outfit he had been wearing. He claimed the lifeboat he had used had capsized when he had crossed over from Green Island. But Edvart and the others had already gauged the trade winds. They knew that if the little boat really had capsized, the waves would have pulled Hans right out to sea. The other castaways had observed that Hans could barely swim.

Edvart confronted Hans, asking where Sanders and Brown were. Hans claimed that Sanders and Brown were still living on the other island, and that the captain had instructed him to tell everyone to leave the atoll on one of the repaired vessels "and take whatever we needed with us except instruments and clothes."

The others told Hans that they would not go anywhere without speaking to the captain first. Privately, Edvart and the others, strongly suspecting foul play, planned to go over to Green Island with Hans and maroon him there if they did not find the captain. With the disappearance of Sanders, Edvart found himself leading the remaining men.

The next morning, Edvart and two additional members of the crew, Martin Neilsen and Charles Clemens, crossed over to Green Island with Hans. While they rowed across, Hans desperately introduced yet another narrative.

"I tell you, boys," Hans said, "the captain killed Brown and I have been keeping him company to prevent him from killing himself." He said Sanders had shot Brown in the back of the head and had forced Hans to dig a grave for the murdered shipmate. He claimed Sanders threatened harm to him if he told anyone the truth. "I was in deathly fear of the skipper." He warned Edvart and the others that they might find a grisly scene and prepared them for the possibility that Sanders might also be dead: "I wouldn't be surprised to find that [the captain] has made [a]way with himself."

They reached the shore of Green Island. As Hans started making his way to where he claimed they might locate the captain, the others all retraced their steps to the boat and rowed back to Sand Island, leaving Hans, the weak swimmer, with no safe way across.

As the others regained the shores of Sand Island, Edvart instructed William to remain on lookout for any signs that Hans was trying to follow. They raced to finish preparing for their departure. The group decided that they would leave some necessities for Hans, who they reasoned would eventually find his way back to the bigger island. Though by now they were certain Hans was a killer, the decision to leave him to his probable death on the atoll weighed heavily. In the hut, they left an axe, food, and other items Hans might use once they were gone.

Realizing he had been abandoned, Hans, as he explained later, was consumed by the horrible truth that "they had left me there to die." He surveyed the surroundings. "I almost went crazy racking my mind for a way to escape," Hans recalled. He collected logs, rolling them across the sand in "the hardest work I ever did." Then he began tearing apart his clothes. He used the scraps to tie the logs together into a makeshift raft. With a piece of wood to use as a paddle, he launched his raft into the water toward Sand Island.

The water became choppy, loosening the lashings. The logs separated until the whole raft fell apart. Hans held on to a single log, staring in terror into the water he knew to be shark infested. His thoughts raced. "My brain burned with hate of my shipmates. At that time I had a mind to kill them all."

All the while, young William remained at his sentry post to watch for any sign of their nemesis. Darkness fell without a trace of Hans, and William returned to the hut to report to Edvart.

Neilsen and Clemens, meanwhile, went to work making sure that Hans would have no way to follow them. They pushed the *Dunnottar* lifeboat out to sea. Then they filled the *Siegel* lifeboat with stones, watching it sink in the lagoon. The only boat left now was the repaired sampan.

At the hut, William prepared dinner for their group, while Edvart sat, charting a sailing route, with a rifle nearby. At the sound of footsteps, he assumed the others had succeeded in sinking the vessels and were now back. But staggering into the doorway of the hut came Hans, naked and dripping wet. He rushed in and grabbed the rifle, pointing it right at Edvart's chest.

"What did you fellows mean by leaving me over there?" Hans demanded.

Edvart jumped up and grabbed the muzzle, forcing its aim away. They struggled until Edvart came out with the rifle.

With the weapon turned on him, Hans seemed oddly resigned to die. He said, "Now you can shoot me."

Instead of shooting Hans, Edvart fired outside as a signal for the other men to come. Together, the castaways prepared to board the sampan. Hans begged not to be left behind, offering to be bound. "Tie me up and take me with you."

Edvart and the others were certain Hans would kill them at sea if they allowed him to come, and it was pointed out to him that he might actually be better off on the island, considering there was a good chance they would drown. They pushed off in the morning.

Hans was alone, except for the birds, which covered almost every inch of the island and could grind down an inhabitant's sanity. Robert Louis Stevenson described what it was like among the atoll's fowl: "It were easier to cross Trafalgar Square upon a day of demonstration than to invade these haunts of sleeping seabirds. The nests sank, and the eggs burst under footing; wings beat in our faces, beaks menaced our eyes, our minds were confounded with the screeching, and the coil spread over the island and mounted high into the air."

"I wandered all day about the island," Hans recounted of his descent into madness, "muttering and cursing, to return to the [hut] at night, then to fall exhausted, yet unable to sleep . . . My mind was a hell."

Chapter 10

THE MONARCH OF ALL

May 1888

THE CASTAWAYS FROM THE *WANDERING MINSTREL*, arriving eight months after Hans was marooned by shipmates, treated him very differently at first than *Siegel*'s crew members had. They needed his help, for one thing.

Hans had transformed in the months he'd been on the island. In the first days after he had been abandoned, he had searched everywhere for the small boat he and the *Siegel* castaways had used to ferry between the two islands of Midway. When the water was clear, he was able to spot the vessel at the bottom of the lagoon, loaded down with stones, revealing his former shipmates' sabotage. In a state of rage and disorientation, Hans stopped sleeping altogether and became ill, succumbing to delirium. He decided on multiple occasions to commit suicide. He suppressed these thoughts by setting himself to specific, tedious tasks, including repairing and improving the hut, and keeping a fire burning from the remaining coal left behind by the American expedition. The work seemed to allow him to sleep, at least, which restored his strength. *The Wrecker*'s version of Hans, a Scandinavian brawler named Goddedaal with a sensitive side, also expressed himself through hard work: "A huge Viking of a man, six feet three and of proportionate mass, strong, sober, industrious, musical, and sentimental." The fictional iteration cherished a pet canary. The actual Hans trained a rat to eat from his hand to relieve loneliness. He would also speak to the rat.

Seven months into his isolation, Hans had cut his own message on the wall of the hut: "Here from September 22, 1886, to January 9, 1888." (September 1886 being when the *Siegel* first anchored at Midway.) He signed this J. A. Jorgensen (J for Johannes, for which Jans or Hans is a diminutive).

When he spoke to the new inhabitants on the island that washed up in lifeboats from the *Minstrel* three weeks after he carved his message, Hans described how he floated over on a raft, which was not untrue, but avoided details.

As for Hans's quick affinity for first mate John Cameron, rumors emerged that the two had sailed together in the past. While John claimed to the others that he did not remember Hans, he could not deny Hans seemed to know him. The lack of records regarding sailors on merchant vessels render it nearly impossible to determine on which prior voyage Hans might have sailed with John, but two possibilities were the steamship *Planter*, which John helmed, and perhaps more likely the *Pomare*, on which Hans could have helped John with the forcible recruitment of labor for Hawaiian plantations.

Whatever the origin of their camaraderie, John relished it. From the time in Honolulu when he stepped aboard the *Minstrel*, John leaned on the loyalty of the burly second mate, George Hanker, to reinforce his open hostility against the Walkers. But George had become increasingly incapacitated on Sand Island.

George, in fact, continued to be emotionally shaken and depressed. The realization, and feelings of guilt, that he might never again see the family he had neglected no doubt weighed on him, contributing to his slipping grasp on reality. His state of mind evokes one of the early meanings in English of the word *castaway*, referring to someone rejected or thrown aside, sometimes for their moral or spiritual failings. George would not stay with the others in the improvised village that extended from the original hut. Instead, he had found a barrel, placed it on its side, and crawled inside headfirst for shelter. "From us all he remained aloof," John observed, "venturing out only in the gloaming to capture some of the myriads of small crabs that infested the beach and then, like one of the animals, scuttling back."

With the loss of George's companionship, the emergence of Hans Jorgensen was serendipitous for John. But if Hans seemed to pledge himself to John, he increasingly alienated all others. They continued to observe startling behavior in Hans, including the angry rants during which he seemed to talk to himself. Reportedly, at the appearance of a full moon, Hans would rave madly, walking up and down the beach. He started arguments with the other castaways. His outbursts felt different to the observers than the rest of the castaways' mental strains, reflecting a darker history than he had revealed.

The situation was increasingly ominous for the Walkers. Elizabeth knew when Frederick, naturally taciturn, was weighed down with anxiety. Hans's alarming behavior loomed over them, and containing him would fall largely on Frederick.

But Hans was not the only one on the island with something to hide. Frederick had brought along a secret objective to the atoll that he could no longer conceal. Back when the *Minstrel* had been waylaid in Honolulu during its near mutiny, Frederick had heard stories about the disaster of the *General Siegel*. The four surviving castaways who previously had been stranded on Midway with Hans—the de facto leader Edvart Olsen, the young German William

Timpe, and shipmates Martin Neilsen and Charles Clemens—had managed to sail their modified sampan to a remote atoll that was part of the Marshall Islands, after a trip of approximately three weeks. Over the course of the subsequent months, Neilsen and Clemens separated from the other two. Remaining in place where they had made land, Edvart and William fortuitously encountered the captain of a New Zealand–based schooner, *Ehukai*. The captain transported them to Jaluit, where they found another schooner, this one headed for Honolulu. Edvart later commented that even if he were paid a fortune, he would never repeat the journey from Midway to the Marshall Islands.

The version of events Frederick Walker heard during that Honolulu stopover was incomplete, but he gathered enough information about what happened to the *Siegel* to understand that there was an accused criminal marooned somewhere on the Midway Atoll—that is, if the man still lived. The Hawaiian authorities seemed content to allow him to perish on a deserted island. While anchored near Honolulu, Frederick actually suggested to the police that an officer could join the crew and, if the schooner stopped at Midway Atoll, could search for the outlaw. The police refused.

But when the *Minstrel* charted course to Midway to fish for sharks, as the *Siegel* had done, Frederick formed a plan to survey the islands for the alleged murderer. The Hawaiian authorities may not have wanted to risk the expense and danger of searching for an outlaw, but the government paid significant rewards for the apprehension of murderers. Frederick was in the unique position to profit from bringing in a fugitive while also securing justice in the name of the murder victims, which included a fellow captain. Besides, Frederick rationalized, no man, however reprehensible, deserved to die alone on an island.

Frederick was by nature compassionate, but he was also quite capable of acting in his own self-interest. He had taken the commission on the *Minstrel* at a financial low point with the hope of a profitable return, but he also knew that sharking revenue was unreliable. Though he projected a positive attitude, pressure mounted on Frederick to try to supplement his income. He had even saved a treasure map drawn by an old sailor from a ship under Frederick's command in 1883. "Captain," the sailor had said, "I can tell you where you can fill the gunroom with bar[s] [of] silver and gold." The man died in a hospital not long after. Frederick continued to think about making that quest on a future voyage. For now, the reported fugitive from *General Siegel* represented another kind of prospective treasure.

The *Minstrel*'s shipwreck had undermined the best-laid plans. As it turned out, when Hans first stepped out of the hut and made his welcoming gesture to the Walkers, he had actually been raising his hands in a reflexive surrender, thinking an armed party had come to arrest him. The fact was, Frederick could not be certain about Hans's identity at first. When the story of the *Siegel* had circulated, Frederick had heard the name of the criminal as Johnson, not Hans or Jorgensen. There had also been confusion in versions of the tale about which island the murderer ended up on, with the two land formations of the atoll, Sand Island and Green Island, transposed in tellings of the *Siegel* shipwreck. Moreover, Hans was not the only sailor unaccounted for in the area. Details of the whereabouts of two of the *Siegel*'s other crew members, Neilsen and Clemens, had not reached Honolulu, and the names of merchant sailors were often confused and changed, sometimes intentionally, as exemplified by the use of the alias Sanders instead of Asberdine by the captain of the *Siegel*. The man who stepped out of the hut could have been any one of at least three people.

By the time Hans's volatile behavior emerged and pointed definitively to the fact that he was indeed the wanted man marooned by his *Siegel* shipmates, Frederick faced a complicated situation.

He had to protect Hans from the *Minstrel*'s increasingly restless crew as the stranger's reserve of food became more tempting every day. Regardless of Hans's criminal past, his safety was still the captain's responsibility. Besides, protecting Hans also meant protecting Frederick's potential reward, the only hope for financial benefit for the Walkers if they survived Sand Island. Fortunately, Frederick still carried his revolver. John Cameron requested a firearm, too, as he had when he advocated for martial law. "Give me a revolver," John demanded. "I'll maintain discipline." It was a reasonable request from a first mate, but something in Frederick made him refuse again, even before the full extent of John's alignment with Hans became clear.

It remains uncertain at what point Frederick told Elizabeth and their sons the truth about searching for Hans and about his intentions to turn Hans over to the authorities. Even once they learned of Hans's crimes, Elizabeth and the children agreed to treat Hans with compassion and the "utmost kindness." Elizabeth, like Frederick, felt a personal responsibility for everyone on the island, including the man she now knew as the "sea convict," struck by the fact that he was "left to die by his ship." John championed the idea of giving Hans the benefit of the doubt and considering him part of their crew. "We could iron [meaning restrain] him if he turned nasty," John insisted. It would become clear that John had selfish motives for supporting Hans.

Over time, Hans's outbursts included threatening comments about his former shipmates on *General Siegel*, declaring, "They'll be a pretty score to settle if ever we meet." The *Minstrel* crew, like Frederick, had been in position to hear whispers of some of the details of the Green Island murders while still anchored in Honolulu. The carving of Hans's name in the hut on Sand Island right near the other engraved message previously signed by William Brown now gained a sinister meaning—the autograph of a murderer juxtaposed with the signature of his victim.

Hans's pattern of violent paranoia from almost two years earlier repeated itself. Back then, he had raged about Glade, after which he targeted his *Siegel* shipmates who had heard his tirades. Now, he was venting about plans to kill the *Siegel* survivors, and once again was faced with witnesses to his death threats. Frederick and the rest of the castaways, upon returning to civilization, could also turn him in for the *Siegel* murders.

George Hanker had special reason to be shocked by Hans's history, as he could easily imagine himself having been one of Hans's victims from the *Siegel*, since George had been part of the crew of the *Siegel* not long before Hans joined it. The terrifying idea could have contributed to the mental strain that had sent George into seclusion in his barrel on Sand Island.

With Hans's true nature exposed, maintaining positivity among the castaways became ever more challenging, starkly contrasting with many of the cultural and spiritual templates for shipwrecks. In the "stranded" narratives ranging from the shipwreck of the apostles in the Bible, to the literary phenomena of *The Swiss Family Robinson* and *Robinson Crusoe*, including their historical analogues such as a "real" Crusoe named Alexander Selkirk, certain themes recurred. Running throughout the examples was the notion of people desiring, consciously or not, to be stranded and isolated in attempts to rediscover faith or versions of paradise or Eden, to strip down to the essentials of life, to escape societal decline and renew the soul. Poet William Cowper imagined Selkirk's perspective: "I am monarch of all I survey, / My right there is none to dispute; / From the center all round to the sea, / I am lord of the fowl and the brute."

But if an uninhabited island had potential as a new beginning, alternatively it represented a possible end, a final version of civilization, an apocalypse, or, as one commentator on the *Minstrel*'s circumstance put it simply, a hell.

The Walkers were living a nightmare version of this: finding your family on a deserted island with a murderer and with no way off.

WHEN THE WALKERS and the other castaways were occupied, Hans found time to get across the lagoon to Green Island unseen. He had several purposes for going there. When he had killed Sanders and Brown, he had concealed the bodies. Now, he constructed rough-hewn wooden crosses and marked their lonely graves with them.

There had been something else Hans aimed to accomplish on Green Island. He wanted to retrieve the rifle and ammunition that Captain Sanders carried when Hans killed him. Hans could hide this without anyone from the *Minstrel* other than his one ally, John, knowing it. John was famously effective with a gun, having set records at Honolulu's Fort Street shooting gallery.

At one point after this, some of the *Minstrel*'s party crossed from Sand Island to Green Island to search for supplies, bringing Hans along, just as Captain Sanders of the *Siegel* had done before them. They came across two mounds of sand with wooden crosses on them with the year 1886 scratched into them. "There was nothing but the rough crosses," a later eyewitness from the era wrote, "the white, desolate sands, the great, boundless ocean and the wild sea birds which perch themselves upon the crosses in ignorance of their sad story." When the discovery was brought up to Hans, he was oddly casual as he explained that those were the graves of his former captain and shipmate from the *Siegel*, though he refused to say anything about how they died.

Still, Hans's grave markers proclaimed a clear message: He had killed one captain who got in his way, he could kill another.

Chapter 11

TEMPESTS

August 1888

FREDDY, HENRY, AND CHARLIE WANTED their parents to regard them as capable as any adults. They rallied around the family's cause to defend themselves against John and Hans, no matter how much Elizabeth and Frederick wished to shield them from danger. As the eldest child, Freddy in particular sought to prove his worth.

Mid-August brought menacing skies, as if to accompany brewing fears of violent conflict. John looked up one evening to observe the changes. "Overhead, a dull and heavy gray; below and extending almost to the horizon from east through north to west, a sickly violet; beneath that lay copper to the dim line where sky and sea met."

That night, the castaways could not sleep, kept awake by the howling gusts. Frederick checked the pressure readings on his barometer, which fell rapidly. He recorded his observations: "When the wind got to the east, the barometer fell one inch and a half; at south another three-tenths." As described by Frederick, "fierce gales of wind howled over [our] small, insecure abiding place, almost blowing [us] into the sea, while the salt spray and showers of sand flying before the tempest stung [our] faces and half-clad bodies."

It was a hurricane. Elizabeth watched from the hut as the "tempest swept over the islands." Sand pelted the rudimentary structures until "a hell of sand" penetrated through the openings, filling eyes, ears, and hair. "We breathed it, swallowed it, gritted between our teeth."

There was another terrifying aspect to the storm: Hans was unaccounted for. He could have been lurking anywhere while the Walkers were trapped in the hut, the walls of which were cracking, as the island was submerged. The storm worsened. Water rose to their waists. If Hans planned to strike against those he believed to be his enemies, he now had them boxed in. There were plenty of reasons for the castaways to be filled with intense fear and suspicion, staring out into the downpour and waiting for an attack.

In fact, Hans was in severe danger, having been outside during the hurricane, and was at risk of being swept to sea. Left on his own, he would have perished, as he collapsed into the ground, where he was being buried under sand. John took it upon himself to retrieve him. He braved the storm, searching until he found the prone figure of the fugitive. "To see was impossible: though the time was forenoon, a dun night enveloped the world," John wrote, "while rain and furious sand stung our faces like needles . . . we crawled . . . on the flats of our bellies, guessing our way." The two men had to lash their clothes to their bodies so they weren't blown off. By the time they reached the village of huts, their faces were bleeding from the sand.

MEANWHILE, OUT ON the open Pacific, a merchant ship called *Brussels* sailed from New South Wales, just as the *Halewood* had, toward Midway. Built twenty-five years earlier in Glasgow, the *Brussels* was an impressive iron clipper that spectators at ports would line up to admire. "Fitted out in splendid style" with fully rigged sails, *Brussels* was strong and powerful, frequently entrusted with large quantities of valuable cargo. But before long, the clipper would go up in flames.

Everyone on board would tell a different story later. According to the first mate, Clarence Caskey, twenty-seven, the captain had pulled him aside to share an underhanded plan: They were going to deliberately wreck *Brussels* at Midway.

Scenarios involving intentional wrecks had occurred throughout history for a number of reasons. The infamous British pirate Edward Teach, known as Blackbeard, may have chosen to run *Queen Anne's Revenge* aground in North Carolina in 1718 in part as a quick way to dispose of a ship with structural problems, and in part to reduce how many sailors would share his bounty. More than a decade earlier, Scottish sailor Alexander Selkirk—whose experiences inspired Daniel Defoe to write the novel *Robinson Crusoe*—asked to be marooned on an island off Chile rather than continue on an unsound ship.

In the fraud scheme as described by Caskey, wrecking *Brussels* would generate insurance money. "The ship would be worth more on the rocks," the captain allegedly explained to the mate, "than anywhere else." The ship was insured for $30,000, and the officers would share this.

According to this version, the captain had identified Midway Atoll as the ideal place to wreck the ship out of sight. The treacherous reef meant it was less likely for insurance firms to question why a ship foundered there, and the location was so remote nobody would go investigate it.

Captain Robert Crosby would later insist that it was not he but

Caskey who was plotting an insurance scheme. By engineering a wreck at Midway, Caskey could offer details to insurers to claim the inevitable reward for information about what happened to the vessel.

Yet another version was that the steward, Charles Bachman, thirty-one, who claimed to also be a journalist, was the plotter. Further variations included Captain Crosby forcing both Caskey and Bachman into a conspiracy against their wills, and Caskey and Bachman plotting together behind Crosby's back.

Regardless of which sailor or sailors on the *Brussels* hatched the Midway wreck plan, it is unclear what they expected to happen afterward. How would they get off the atoll after running aground? They may have communicated to a coconspirator that another vessel should sail for Midway to retrieve them. Another possibility is that *Brussels* carried an unusually strong, well-equipped lifeboat in which the plotter or plotters could feel confident, wisely or not.

Specifics aside, they could not know that Midway already hosted castaways, in the throes of their own highly charged dramas.

Even if the version of the narrative told by first mate Caskey was true about Captain Crosby originating the plot, Caskey acknowledged temptation. In fact, as they came closer to Midway, Caskey apparently negotiated with the captain to determine his share of the insurance fraud. Failing to come to terms, the first mate refused to join the plan.

The Midway scheme may well have brought *Brussels* close enough to spot signals of smoke from the castaways, which the Walkers kept burning whenever possible. The *Brussels* crew would have been shocked at this development, which presented an unexpected risk to their machinations in the form of witnesses. They could not have guessed the power struggle unfolding on those shores. Perhaps the plotters also thought twice about their prospects of even surviving a deliberate wreck at Midway as they came closer to the jagged reefs.

Whatever their doubts, they decided not to stage a wreck, changing course and sailing past Midway, continuing the long voyage to San Francisco.

An Episcopal bishop arriving in San Francisco's "noble harbor" years earlier observed a tendency toward big risks and big rewards for those making port there: "Some are coming to retrieve broken fortunes, and, instead of reaping the golden harvest, how many will find a grave in the already crowded cemetery of San Francisco— dying, 'strangers in a strange land.'" Some of *Brussels'* officers certainly qualified as reckless fortune seekers. The *Brussels* anchored at San Francisco with a new plan in mind to obtain the insurance money: setting the ship on fire.

In the middle of the next night, excitement spread through the wharves. Ringing bells and flashing lights came from the water. Rumors had started that a recently arrived British ship called *Ravenswood* had been on the verge of a mutiny, and the commotion was thought to be connected to that. Two police officers and a waterfront reporter from the San Francisco *Examiner* at the harbor climbed into a boat and headed in the direction of the noise. They discovered the source was not the *Ravenswood* but the *Brussels*, and the cause was not mutiny but conflagration.

Caskey would later describe being woken up from his bunk by shouts of "fire" from the small number of crew members on board. From the open hatchway, "the fire seemed to leap to the heavens," he said.

"Flames were issuing through the hatchway at an alarming rate," wrote the reporter, "and threatened every minute to communicate with the rigging . . . The vessel was laden with whale and coal oil, petroleum and other combustibles, and it was feared every minute that an explosion would occur." A firefighting boat arrived, pumping five thousand gallons of water a minute through five streams. Caskey, meanwhile, would describe how he heroically directed the crew to seal the hatch door.

Once the fire was extinguished, investigators gathered details and insurance companies began investigating through underwriters, as well as requesting assistance from the San Francisco Police Department. Taking the lead was the city's top detective, Isaiah Lees, fifty-seven, who had relocated to Northern California years earlier from England.

Reviewing the circumstances of the *Brussels* fire, Lees assigned two of his detectives to shadow Caskey and the steward, Charles Bachman, who spent time together in a saloon onshore where the police noted "girl waiters are employed." The detectives found that Bachman had given a gold watch to one of the waitresses despite the fact that he had told the insurance company that he lost the watch in the fire. They also found other gifts given to waitresses that Captain Crosby now reported as stolen from his stateroom before the fire. Investigators found no petroleum in the cargo, which is what Caskey had claimed spontaneously combusted.

These discoveries led to the arrests of the first mate and steward. In a twist, Caskey accused the steward, Bachman, of using psychic manipulation projected through the glare of his dark hazel eyes. "He had perfect control over me. I was entirely under his mesmeric influence. . . . His power over me is terrible and hellish." The allegations attempted to evoke anti-Semitic feelings against Bachman, though it was unclear whether the steward was even Jewish.

Both Caskey and Bachman were convicted of arson. Their stories changed so many times that the records of the case turned out garbled. Captain Crosby was never charged, but the crew members' confessions often included the captain giving them instructions to carry out the fraud. Chief Detective Lees was generally celebrated as "remarkable for his cleverness in ferreting out criminals and frustrating their plans." However, critics of Lees alleged that his greatest skill was taking credit, particularly from his longtime rival, private detective Harry Morse. "At any time the Chief has a big case he cannot handle," Morse wryly noted to a reporter, "my services are at his disposal."

The critique that Lees flattened complications in service of resolving his cases applied to locking up *Brussels'* first mate and steward without fully exploring Captain Crosby's role in the conspiracy. Reporters at the scene determined that Crosby was "impossible to find" on the night of the fire, and police were initially vocal about their certainty that he took part in the scheme. Crosby quickly left the country on another vessel, claiming he was ordered away by the owners of *Brussels*, while Caskey testified that the detectives "bullied him into making a deposition exonerating Captain Crosby." Detective Dan Coffey, whom Lees had assigned to gather key evidence against the two indicted crew members, later exhibited signs of serious mental illness, and "would burst out crying like a child" when spoken to sternly by Lees.

The conspirators must have had second thoughts about whether they would have been better off if only they had carried out their initial scheme to scuttle the ship at Midway, where evidence would have been washed away by the sea. Regardless, now that they were in San Francisco, they still had ample opportunity to report any signs of life they had seen when they approached Midway.

However, Captain Crosby had reason to stay quiet. If he informed the press or the authorities about a possible wreck at Midway, an imminent rescue of the *Minstrel* castaways may have taken place, but questions would have been raised about *Brussels'* strange detour, ultimately implicating Crosby in the criminal conspiracy. Caskey and Bachman, sitting in their prison cells with fifteen-year and thirty-five-year sentences, respectively, also had no incentives to speak up. Whatever intelligence they possessed about Midway, none of the former *Brussels* crew shared it.

BACK ON SAND Island, after sixteen relentless hours of rain and violent gusts, the hurricane dissipated. The scant vegetation had been torn up, looking like it had been "scorched as by fire." During the height of that storm, parts of the islands were entirely washed over, a reminder of how precarious the land was. Nature had reshaped the terrain on Sand Island overnight, with sand shifted into new elevations and ditches. When the castaways emerged from the shelters, "the succeeding hot, blistering sun completed their misery" by burning into the cuts that had been left across their bodies from the sand. The Walkers and the others began repairing their so-called village from the damage of the storm.

John and Hans moved to Green Island, where they "took themselves apart from the rest of us," as the Walker sons observed, suspecting such a removal precipitated an escalation of trouble. John also convinced his formerly close ally, George Hanker, to come with them. George, however, insisted on bringing his barrel, which would not fit in the rickety lifeboat. They dismantled it into staves (or planks) and hoops. Two of the staves fell into the water as the trio crossed the miles-wide lagoon. Once on land again, George could not reconstruct the entire barrel, so he created a sort of kennel with its components.

In addition to the hostility they had generated from the rest of the group, John and Hans had another reason to exile themselves. Part of the *Siegel* wreckage remained lodged in sight near the beach—with an apparent hidden fortune inside.

Before joining the *Siegel*, when Hans served on the German sugar transport vessel *Apollo*, transportation of sugar had long been used by smugglers as a cover for carrying opium between Hawaii and the United States. Opium itself was not entirely illegal in the two countries, but use was restricted to medicinal and pharmaceutical purposes. Opium passed through customs, where importers had to pay such high import tariffs that black markets grew in both places, requiring a veritable army (and navy) of smugglers.

Once Hans found opium on *Apollo*, he may have deserted the vessel in order to abscond with some for his own profit. *Apollo*'s captain could not acknowledge storing unlawful cargo, which limited his options for recourse with the German consulate, but he could readily demand Hans's arrest as a deserter. Still, the German consul had not acted decisively enough to prevent Hans from joining the *Siegel*.

The unusual circumstances of the *Siegel*—the loose supervision by owners planning to sell it, the shorthanded crew—generated ideal conditions for Hans to smuggle stolen opium on board, until the wreck at Midway intervened. The prospect that he had brought contraband shed new light on the murders Hans committed at Midway. Perhaps Captain Sanders or William Brown had found out about his smuggling, or maybe Hans tried to coerce them to help him recover opium from the wreckage that he could not retrieve by himself. If the old sailor Peter Larkin had known about the opium, this might have motivated Hans's apparent plot to poison him while he recovered from the dynamite accident. Now that Hans joined forces with John Cameron, more possibilities presented themselves. With the *Siegel* wreckage caught in the reef, Hans needed help from a second diver to reach a hidden compartment inside what remained of the *Siegel*. John, too, had experience with opium smuggling. His earlier refusal to acknowledge his previous acquaintance with Hans had concealed their mutual history in the opium trade from the other castaways.

The Wrecker dramatized a search for opium inside wreckage resembling what John and Hans faced. "The hold was ceiled [meaning lined] throughout; a part, where perhaps some delicate cargo was once stored, had been lined, in addition, with inch boards; and between every beam there was a moveable panel into the bilge. Any of these, the bulkheads of the cabins, the very timbers of the hull itself, might be the place of hiding." Stevenson and Osbourne's fictional salvagers, or "wreckers," worked in unison to conduct their intense search, just as Hans and John would do. "It was therefore necessary to demolish, as we proceeded, a great part of the ship's inner skin and fittings, and to auscultate what remained, like a doctor sounding for a lung disease. Upon the return, from any beam or bulkhead, of a flat or doubtful sound, we must up axe and hew into the timber: a violent and—from the amount of dry rot in the wreck—a mortifying exercise." The wreckers' reward was a series of small tin boxes of opium.

Another advantage of going to Green Island—aside from the possibility of finding the opium—was that John and Hans had free rein to operate away from prying eyes. Using Hans's expertise, they built grass huts. They also had begun constructing another boat using materials brought over from Sand Island as well as timber and scraps found on Green Island. Both men had already displayed their capacities for ruthlessness, and both had reason not only to separate from the others, but also to wish to eliminate the other castaways—or at least the most influential among them.

Midway represented a chance for John and Hans to find new starts. John's reputation already having been tainted by the death of the young woman on the *Planter,* he did not want anybody repeating stories of his hesitation as the *Minstrel* broke apart, which could well have cost more lives. Hans's stakes were arguably even bigger, needing to escape the consequences of his crimes against his *General Siegel* shipmates. While working on salvaging and stripping wreckage and thinking about the violence that could still be necessary, John trembled to consider Hans's bloody history on the same ground they walked upon. "Did ghosts watch us at work," John wondered, "one-handed Peter Larkin, Brown with a bullet hole in his head, the lost [Captain Sanders]?"

The Walkers did not possess the telltale clues of a murder plot Edvart Olsen had found when he'd discovered Hans's falsified diary entry about missing castaways. Still, the violent undercurrents coming from Green Island became impossible to ignore, especially since the exiles had access to the weapons cache of the murdered *Siegel* captain. Details of what happened next remain scarce. According to one report, when Frederick came to John and Hans's encampment on Green Island to check on them, Hans shot at him, but missed.

Other evidence suggests Hans orchestrated an ambush on the Walker children. In the murderer's logic, killing one or more of the sons would break the spirit of Frederick and Elizabeth, thus leaving the rest of the castaways in a collective state of emotional chaos. Hans slipped away from Green Island in the lifeboat in which they had come over, paddling toward Sand Island. The semidelirious George Hanker discovered Hans's plan after overhearing a conversation between Hans and John. Over on Sand Island, Hans concealed himself with his rifle, waiting for his opportunity. The Walker boys tended to stay close to each other, with Jessie, despite her diminutive size, fiercely guarding them. If Hans managed to shoot one of the boys, the other two would rush over to help, giving Hans clear shots at them too. But George, who had caught wind of the plan, floated or swam across the lagoon in time to find Hans. George drove Hans away and the Walkers escaped unharmed.

Still, Frederick had to face the fact that his attempted shortcut to financial gain had trapped his family with a killer. On top of this, his first mate, albeit one thrust upon him at the last minute, had betrayed him and the whole ship. "I never in my life found two such unmitigated liars," fumed Frederick about the mate and the fugitive.

Back on Green Island, John and Hans had to anticipate retaliation for the foiled attack on the Walkers. Working feverishly before winter brought another bout of rough weather, John and Hans lined the bottom of the vessel with planks, and improvised staples from wire rigging recovered from the wreck. They created a sail from leftover portions of the *Minstrel*'s materials, facing setbacks every time a fresh rainfall soaked them.

Colder weather set in quickly, with the number of birds available to hunt diminishing. The two exiles smoked mullet and pickled birds so that they would last longer for their voyage. John and Hans rationed their eggs, setting aside as much food as possible for the trip. The birds at this time of year could only be hunted at night. "Usually we stalked them before dawn," John noted, "when we rushed upon them and killed them with clubs."

On Sand Island, a heightened sense developed that the Walkers would be the hunted ones again. With second mate George Hanker redeemed but mentally unwell, the structure of authority on the atoll, already fragile, degraded. Besides, before his breakdown George had been loyal to John Cameron, and increased strife tested other allegiances. John boasted skills of persuasion he had cultivated for years while manipulating recruits and officials during his blackbirding expeditions. He now managed to enlist young Mouwen, or Moses, the Chinese sailor John had coerced into salvaging his tobacco from the sinking *Minstrel*. No doubt Mouwen's limited English appealed to John, reducing the chances Mouwen would overhear secrets. One part of his plan John did not want anyone to suspect was that he would procure an essential tool—Frederick's compass—by "force to gain my ends."

The two islands represented two different types of settlements or, in a grander vision of the situation, civilizations. The Walkers and the fourteen castaways on Sand Island cooperated and coordinated with each other to provide and allocate enough resources to sustain themselves. The alternative settlement on Green Island found two leaders who used threats of force and violence to strengthen their positions, taking advantage of the challenges in mental health and language, respectively, in George and Mouwen in order to exert control over them.

Frederick confided in Elizabeth: he was going to shoot Hans. After the ambush by Hans of the Walker sons, lines had been crossed. To protect his family, Frederick had to commit the same act that had made Hans into an outlaw—murder.

They could not know how close USS *Juniata* had come to landing at Midway seven months earlier, with its 21 officers and 179 crew members, including highly skilled marines (soldiers trained to serve on a ship and at its landings) armed with rifles, a howitzer (similar to a cannon), and a Gatling gun (an early machine gun). But no such force came to save them, and Frederick felt he had no choice. If he crossed stealthily to Green Island, he calculated, he would have the advantage of vegetation in which to hide himself while waiting for his best chance to shoot. The boys, by this time targets themselves, would be particularly anxious to step up and help eliminate Hans. Frederick prepared on multiple occasions for a showdown with Hans, but each time Elizabeth stopped him.

Elizabeth ultimately had been in favor of Frederick's original plan—to bring Hans back to Honolulu authorities—which would have spared the fugitive from certain death and allowed the Walkers to collect a much-needed reward. But that had gone terribly wrong and was arguably misguided from the beginning. She also knew that, if Frederick tried and failed to kill Hans, the image of the wood crosses atop bodies of Hans's victims could presage her own family's future.

Frederick's analysis involved assessing how many weapons and crew members they had at their disposal compared to what John and Hans could count on. But Elizabeth could not advocate killing Hans. With her husband's judgment thrown into question, she filled the leadership vacuum. In her view, the Walkers could not protect themselves from both the outlaw and the treacherous first mate, who was arguably even more devious.

Her more sophisticated idea was counterintuitive: instead of trying to kill Hans, she suggested they offer to supply John and Hans with some of the precious resources with which to leave the island. This would prevent the two exiles from harming the Walkers and the other castaways. Moreover, in signaling permission for the exiles to take the boat and a share of supplies, they could extract a vow from them to report their whereabouts to the authorities once they reached safety. The plan pivoted on the rational supposition that, if the breakaway faction made it to land, they would want money, and would report the castaways' whereabouts in return for the rewards posted by now by insurers.

The looming showdown transformed into a negotiation. John

demanded that they be allowed to bring the larger of two compasses, but Frederick agreed to part with a smaller, silver-bound family heirloom compass, with the request that it be returned once they all were safe. They also gave John's faction water casks holding a total of 150 gallons.

On October 13, 1888, John, Hans, and Mouwen boarded the boat. Frederick handed John a letter he wrote to be given to the authorities.

Watching Hans cast off from shore left the Walkers to contemplate how their perception of one person could be overturned so dramatically. Hans had been seen as a savior, then an existential threat, now back again to a potential savior. Such a fascination with an individual's duality lined up with the times. Robert Louis Stevenson—who would later investigate the happenings at Midway—had mesmerized readers with *Strange Case of Dr. Jekyll and Mr. Hyde*'s account of evil lurking within a respectable exterior, a horrifying notion to the reading public. In fact, at the same time the castaways struggled on Sand Island, citizens of London and beyond had become riveted and frightened by what would be known as the Whitechapel or Jack the Ripper murders. Part of the fear arose from the notion that the murderer could be an ordinary, mild-mannered person strolling the streets of London—which linked back to the dual nature of Jekyll and Hyde. Just the week before the Walkers gathered at the beach to see off their tormentors, a member of the public insisted to the Metropolitan Police in London that thespian Richard Mansfield, who was portraying Jekyll and Hyde in a play, must be the mysterious Jack the Ripper himself, so well did he express both good and evil.

After John, Hans, and Mouwen pushed off from shore, the others discovered that the exiles had taken far more than their share of supplies. They had stolen "all of the tools . . . saved from the wreck" from the main encampment on Sand Island. Whether they would make it to safety or not, John Cameron reinvented himself by thrusting himself back into the role that had eluded him for so long: he was a captain again, and would let nothing get in the way of keeping that position.

Chapter 12

LIGHTS

Fall 1888

JOHN AND HANS'S THEFT OF supplies aside, the Walkers felt profound relief to have their antagonists off the island. Tensions lessened, though their island imprisonment, as Elizabeth called it, continued without abatement. The second mate George remained despondent and inconsolable in his so-called kennel. Elizabeth pleaded with him to keep faith as she endeavored to do. She urged him to believe that help would come. It had to.

In the day-to-day repetition of tasks, they tried to stay positive. At its most beautiful, Sand Island evoked one of the Hawaiian words used for the Midway Atoll, *Kauihelani*, or the backbone of heaven. On a clear day, tiny pieces of coral gleamed brilliantly across the perfectly white sand of the "glittering beaches." When darkness fell, as Robert Louis Stevenson and Lloyd Osbourne wrote of the island's quiet enchantment in *The Wrecker*, "[they] saw the hollow of the night, the bare, bright face of the sea, and the stars regarding them; and the voices died in their bosoms at the spectacle."

In the weeks following the departure of the first mate and the fugitive murderer, Frederick prayed every night for "the sight of a sail."

One evening in mid-November, they spotted a red light on the horizon. It could have been a momentary reflection, maybe an optical distortion. Then it turned green. On many ships at the time, a red light was required to be used on the port side (left, when aboard and facing forward), and a green light on the starboard side (right, when aboard facing forward), to communicate right of way on a clear, dark night from at least two miles away. Judging from the movement and speed, they suspected the lights belonged to a steam-powered vessel.

The castaways had kept track of dates, so they surmised that by now, approximately three weeks since the lifeboat's departure, John, Hans, and Mouwen must have made it to Honolulu and had notified the authorities, who had come to save them. Their archenemies, if only to collect rewards, may have become their liberators after all.

Elizabeth described how they jumped into action, with "all signals and their every effort," at the green and red glow from the unknown steamship. They built fires at the highest elevations on their low-lying island.

THOSE LIGHTS THE castaways saw headed for them belonged to a giant four-masted sailing steamship owned by the Pacific Mail Steamship Company. The *City of Peking* ran routes between Hong Kong and San Francisco, with a stop in between in Yokohama. When constructed the previous decade, the 423-foot-long *Peking* (along with a sister ship) was considered the second biggest vessel in the world, and possibly the largest ever to be built in the United States. As a titan of the ocean, *Peking* had saved more than one ship in peril. Once, the steamer towed a stranded schooner called *Florence*, after which observers marveled that the extra weight "did not make any difference in the speed of the ship."

At *Peking*'s lowest level of what was called an "Ocean Palace," the "stokers," usually Chinese workers, fed coal into three furnaces, which powered ten boilers, generating steam power to a pair of 4,500-horsepower engines, which in turn kept the ship's propeller in motion. The *Peking* in more ways than one provided a contrast to both schooners that had sailed the same waters into Midway Atoll the last couple of years. While *General Siegel* had approached the reefs with fewer than ten people on board, the *Peking*'s crew alone could number 160, depending on availability and the needs of a voyage. Nor could *Peking* come and go without much public attention the way a merchant schooner such as the *Minstrel* did. The steamship's routes, arrivals, and departures were tracked in detail by the press, while curious bystanders, including world leaders, waited at harbors to behold the vessel.

At the helm of *Peking* was forty-eight-year-old William Brownell Seabury. It was an apt surname for a man who came from a long line of ship's captains. In fact, Seabury was such a well-known name in shipping circles that a "Captain Seabury" was one of the names in a hoax letter about the capture of a sea serpent that generated international headlines in 1852. Captain Seabury appeared to have been related to the Captain Brownell who helmed the *Ke Au Hou* from Honolulu that had been charged to look for the *General Siegel*, then had sailed right by Midway. Raised in the busy port of New Bedford, Massachusetts, Seabury started by working on vessels involved in the sugar trade. He was steady in crisis, as he proved in 1879 when he steered a burning steamer, *Alaska*, to safety. One reporter who watched Seabury in action called him "a class by himself when it comes to handling a steamer," and wrote that the ship moved "as quickly and easily as if she had been running on a railroad track." Seabury said, "I never have been wrecked," adding the qualifier, "not since I have been steamshipping," because even the steadiest commander had experienced troubles earlier in his career in riskier vessels. Seabury prided himself on the fact that he had not taken a vacation in more than twenty years.

The *Peking* carried cargo ranging from flour, ginseng, apples, and beans, to machinery, opium (being legally imported), codfish, leather, and gold and silver bars entrusted to Pacific Mail by banks in California and China. The steamer accommodated 1,600 passengers, only one tenth of whom were luxurious first-class-ticket holders such as the young British journalist Rudyard Kipling, who sailed on the *Peking* in the same era. The majority of the *Peking* passengers—many of whom came from China—traveled in cramped, unpleasant quarters called steerage.

On its most recent arrival to San Francisco, *Peking*'s steerage passengers ended up as the focal point of a legal dispute. In earlier decades, the United States had facilitated the immigration of Chinese citizens to work as manual laborers on the railroad, on farms, and in factories. Anti-Chinese sentiment grew alongside immigrants' importance to the economy, and resulted in the passage of the Chinese Exclusion Act in 1882. Weeks before *Peking* steamed into San Francisco harbor in October 1888, another piece of complex and confusing national legislation, the Scott Act, further restricted entry of Chinese travelers into the United States. Secretary of State Thomas Bayard frowned upon the legislation, probably motivated by his concerns about disruption to trade rather than about injustice toward Chinese immigrants and Chinese Americans. He encouraged President Cleveland to withhold support, but in a highly polarized political climate and under mounting pressure, Cleveland signed it into law. Bayard admitted helplessness. "The Scott Act was suddenly and without notice brought forward in Congress and passed with a unanimity in both houses which palpably rendered an interposition by a veto of the Executive, wholly futile."

With *Peking*'s status as one of the first vessels arriving under the new law, authorities in San Francisco refused to allow Seabury's Chinese passengers to disembark, even citizens. After legal battles, some passengers, including a Chinese boy born in the United States, received writs of habeas corpus from the court that allowed them to leave the ship, while being demeaned by anti-immigrant commentators as "subjects of the Celestial Emperor" and "Mongolians." Several of those same passengers were forcibly returned to the vessel once the courts claimed they failed to verify their identities.

Even the Chinese crew members of *Peking* were not exempt from the restrictions, meaning many were stuck onboard at the harbor for nearly two weeks. Captain Seabury felt protective of those hard-working sailors and laborers who made the *Peking* run, publicly praising his Chinese crew as more trustworthy and reliable than European sailors, who tended to get into drunken brawls.

A sense of how different it was to experience the *Peking* in the comfort of passenger accommodations was documented by Rudyard Kipling. In contrast to the strain of labor, Kipling contended with boredom. He recalled how he would seek out novelty during the passage, which included looking for shipwrecks. "The monotony of the sea is paralyzing. We have passed the wreck of a little sealing-schooner lying bottom up and covered with gulls. She weltered by in the chill dawn, unlovely as the corpse of a man; and the wild birds piped thinly at us as they steered her across the surges."

Captain Seabury also wanted to break up the languor of the long journey between Hong Kong and San Francisco. In fact, when the Walkers spotted the *Peking*'s red and green lights, Seabury was not steering to Midway because of any communication from the castaways John Cameron and Hans Jorgensen, nor from any sightings reported by remorseful fraudsters from the *Brussels* languishing behind bars in San Francisco. Seabury had no reason at all to suspect that *Wandering Minstrel* ended up on Sand Island. Instead, curiosity led him there. He discussed with a pastor traveling on *Peking* why a place such as Midway Atoll, so "seldom visited by mariners," intrigued him, which came down to a simple sentiment: "In the midst of the awful solitude of the Pacific rest these lonely islands." Seabury decided to go that way because he sometimes imagined that he could "find persons there who have been shipwrecked."

Seabury knew steering through the reef could be treacherous, which was why the American government, influenced by Seabury's employer Pacific Mail, had tried at such great cost twenty years earlier to blast a safer passage. Fortunately, Seabury was known by observers to have "rare judgment and the nerve to back it up," to the point where "his subordinates worship him and even his enemies respect him." Not only could he expertly steer the Pacific Mail steamers, but he knew these vessels from the inside out, personally supervising the construction of the firm's latest ship. Still, the usual potential pitfalls that could come from entering Midway multiplied on this voyage. Seabury had responsibility for more than one hundred Chinese passengers making a return across the Pacific after being refused entry in San Francisco. Being delayed at Midway would be unfair to them, and the possibility of becoming stuck on the reef, even temporarily, could be detrimental to those passengers' health. Seabury decided it was too risky, "afraid of the results," as he later said, "if he came nearer."

WITH THE EXCITEMENT on Sand Island from the beautiful sight of the red and green lights, Elizabeth would have searched for another positive sign, as a castaway from USS *Saginaw* had, for "a shadow . . . developing into a long and well-marked line of smoke" of the steamship approaching. Instead, the Walkers watched in despair as the lights from the steamer gradually grew dimmer until they disappeared, and with it their best chance at deliverance. No amount of shouting or crying could bring it back.

Chapter 13

MASTER OF MISDIRECTION

October 15, 1888

NOBODY HAD HEARD FROM THE breakaway faction—John Cameron, Hans Jorgensen, and Mouwen—because they remained adrift at sea. At first, John was delighted to be anywhere other than Midway after being stranded for nine months. He felt emboldened and prepared because of their survival on the atoll: "I was never hardier, never more fit to endure any privation."

They aimed to sail approximately fifty miles away to Ocean Island, where they knew that the Hawaiian government had constructed its "survival shed" during the earlier search for the castaways from *Dunnottar Castle*. If they could replenish their strength there, they would sail on to Honolulu. Everything felt auspicious, including clear weather and ocean currents pushing them in the direction they wanted to go. With the first night came downpours, but they came prepared. They propped up a piece of canvas on curved supports.

By the middle of the second day at sea, by John's rough calculations, they should have sighted Ocean Island. But their hearts dropped when all they saw was ocean. Their limitations became clearer. When the wrecked *Saginaw* years earlier sent a lifeboat out to sea with five castaways, one of the officers had created a chart of the section of the Pacific, and another officer drafted nautical almanac tables to use for navigation. But John and his crew of two had only a compass and a sextant, without a chronometer or other navigational tools. They were relying partially on guesswork, and even a slight inaccuracy in their estimated course would render a low-lying atoll invisible.

The fact was, if they had reached their intended destination, they would have found that Ocean Island's water tanks and shed had been knocked down by the wind. The food cache had also been taken by a passing ship. But John, Hans, and Mouwen would not have to face the disappointment of discovering all of that, because they never found Ocean Island.

Meanwhile, dark clouds amassed, bringing heavier rains and swells that made a mockery of their canvas shelter. With the three sailors seasick and struggling, they altered the target of their voyage. They followed their compass south with hopes of moving toward the Marshall Islands. This brought calmer water in warmer latitudes, but as the days passed fresh doubts accumulated about the value of John's latest calculations, carried out on loose sheets of paper.

Rain fell and strong gales blew at regular intervals. They fished with a spear brought for the purpose, but their movements caused the unsteady craft to spin off the course. Few fish came near the vessel, and a couple of weeks in, Hans stole food they had been rationing, leading to a confrontation between him and John that nearly turned physical.

In case they ran out of food, John had a secret survival plan that he had carefully thought out. He knew that "a man starving to death at sea doesn't die once: he dies twenty times." Before stepping foot on the vessel on Sand Island, he had plotted how he would murder his companions in order to eat them. In fact, he acknowledged having chosen Mouwen for the voyage because the boy was "young, fat, and probably tender." John's plan was to first murder Mouwen, then eat him, then kill Hans when he needed more food. John called the cannibalism plan "murder for self-preservation."

After two weeks, John assessed that they were 720 miles south of Midway—less than half the distance to the Marshall Islands, meaning they would have to survive at least three more weeks if not a month or more. Illness beset the three passengers, leaving the boat drifting with nobody at the tiller. With more rain, they struggled to bail water out of the vessel. One particular thunderstorm draped them in darkness for five consecutive hours. "Eternal damp was rotting our clothes," observed John. "Our blankets were tearing like paper." There had not been a hint of land or any trace of another sail.

Thirty-eight days in, they spotted coconuts floating in the ocean, suggesting they were closer to land. Hans was ranting and hallucinating and talking of welcoming death. Mouwen cowered in fear of the fugitive. John, concerned about a potential assault by Hans, kept a club at hand, as well as other hidden weapons he had brought for his murder-for-cannibalism fallback.

Forty-three days and 2,700 miles (some of it in circles) had passed when Mouwen, on lookout, cried out, "Land, ho!" They had reached the Marshall Islands.

At Mili (or Mille) Atoll, one of the Marshalls, islanders towed them in. Mili's inhabitants had experience with unexpected visitors, sometimes with dramatic results. One example occurred after a Nantucket-based whaler, *Globe*, became the scene of a violent mutiny in 1824. Landing at Mili, the sailors pressured islanders to choose sides between competing factions of mutineers. Conflicts erupted, with the islanders killing those mutineers who threatened their community, and sheltering the others, who remained on Mili and another nearby island for years afterward.

As the latest visitors were ushered onto the island, John Cameron had to prepare himself for a range of possible receptions. The islanders taking them to safety came from some of the villages where years before John had forcibly recruited and kidnapped laborers to be sent to Hawaiian plantations. However, the passage of time and changes in the population and in John's appearance protected him. In fact, the islanders showered them with hospitality. They gave them cigars and, after a night's sleep, a feast that included "pigs, chickens, ducks, and fish, wrapped in large green leaves and baked in underground pits; rice boiled in coconut milk; cakes made of pandanus fruit."

Islanders repeatedly asked the strangers for the story of their voyage to Mili, which John presented as near-miraculous, not without some justification considering the odds of their survival. John also exhibited his tendency for shaping narratives beyond recognition. According to his rendition, he had been the hero on the voyage many times over, while the other two sailors blundered. While at sea, he had also delivered, at least in his imaginative telling, long monologues to lecture, chastise, and inspire the weaker-willed Hans, while the outlaw sobbed at the bottom of their craft. John, of course, failed to mention his plans to slaughter Hans and Mouwen and eat them. To his campaign of reinvention that had begun back on Sand Island, he now added benevolent savior.

The three found passage to Jaluit, another atoll in the Marshalls, where Mouwen split off from the other two. For Hans, Jaluit was an ideal starting point in his quest for revenge against the *Siegel* crew who had marooned him. Edvart Olsen and William Timpe had also been in Jaluit, and Hans could pick up their trail from there.

John, meanwhile, had to decide what to say about the *Wandering Minstrel*—if anything.

Enough time had passed for the owner and insurers of the *Wandering Minstrel* to realize something was very wrong and that the ship must have sunk or wrecked. Even so, no particular protocols were in place to initiate a search. As happened when *General Siegel* disappeared, the various nationalities of owners, investors, the captain, and the sailors left no single authority to assume responsibility. In one telling comment that reflected uncertainty, a newspaper in Peterhead, Scotland, where the *Minstrel* was built, inaccurately speculated that because the schooner had sailed out of Hong Kong it probably had a Chinese captain and crew. The *Minstrel*'s owner of record, a sea captain named John Pitman, did not have resources to fund an effort, nor did the scattered investors in the sharking consortium that commissioned Frederick's voyage. In Honolulu, Commissioner Wodehouse, who had overseen the brief intervention with disgruntled officers of the *Minstrel* that had led to the addition of John as first mate, had already been embarrassed when chartering *Waialeale* to rescue the *Dunnottar Castle* castaways. He did not want to repeat that experience. If there was any possibility of marshaling a rescue party, John and Hans would have to inspire one.

Settling into Jaluit, John and Hans met with two consular representatives, one American, E. M. Morgan, and one German, Franz Leopold Sonnenschein, whose position was styled as Imperial Commissioner. Sonnenschein assumed authority over all newcomers since Germany had declared the Marshall Islands a protectorate. As the first officials with an opportunity to investigate John and

Hans, they might have asked a few key questions that could have led to hints about the fate of *Wandering Minstrel*. As a start, Commissioner Sonnenschein could have obtained information from Honolulu about Hans deserting a German vessel, which in turn would have led him to the stories of murders from the *General Siegel* survivors. However, neither official had incentive to make such efforts. Morgan worked for the San Francisco–based trading firm Crawford & Company, serving the lucrative blackbirding market, which relied on disreputable sailors with questionable histories. He was not in the habit of scrutinizing strangers. Meanwhile, Sonnenschein was primarily interested in the new arrivals paying taxes.

The questions the islands' consular representatives did think to pose led nowhere. After all, John mastered misdirection and had experience fabricating facts about a ship for his profit and other personal benefits. He now prepared to invent a new tale; to do so, he needed a false ship name and shipwreck, and had to make the details believable.

When asked, John and Hans, who had coordinated their stories, reportedly said they were survivors of the *Thistle*, and apparently indicated that they had wrecked on Morrell Island, described in navigational texts as "a small, low island, nearly level with the water, and about four miles in circumference, with reefs extending from it for about 15 miles to the west, and about 30 miles in a SSE direction," making for a treacherous approach.

The ship name they gave to Morgan was a sly joke by Scottish John, as the thistle was the national flower of Scotland, a way for John to add a secret signature to his saga. Morrell Island was another crafty detail. The locale was included on many maps since Captain Benjamin Morrell had first discovered the isle in 1825. But different maps indicated different sets of latitudes and longitudes, and voyagers who sought it out had trouble locating it, leading to exciting speculation about the "disappearance of a Pacific Island." In fact, the island never existed at all, one of many fabrications by Captain Morrell, but at the time few had realized this. Morrell Island provided the perfect setting for John and Hans's own fabricated shipwreck, a counterfeit of a counterfeit that could not be traced.

The two most important pieces of cargo John had to deal with were the letter Frederick Walker entrusted to him pleading for help, and John's share of the opium salvaged with Hans from the wreck of the *General Siegel*. But the opium comprised part of a long-term plan for which he would have to wait to turn a profit. As for the letter, the time had come for a decision. Would John share it with authorities, as Elizabeth Walker had hoped, even if motivated only by personal gain? Apparently not. The letter disappeared and was never seen by Consul Morgan or Commissioner Sonnenschein—or anyone else. Instead, John likely burned or otherwise disposed of it on Jaluit.

But about one thing, at least, Elizabeth Walker was correct: Despite the *Thistle* fabrication and the destruction of the all-important letter, John was not able to resist the temptation of chasing reward money. The *Minstrel*, like many vessels, had been covered for loss by multiple insurance firms, several of which offered rewards in order to try to reduce their liabilities. Policies for the *Minstrel* had been issued by firms in China, England, Scotland, Hawaii, and the United States.

The first reward John could have chosen to pursue was being offered in Hawaii. A prominent insurance agent in Honolulu named Charles Otto Berger posted an "Information Wanted" notice for the *Minstrel*. But for John to return to Honolulu to try to claim that reward would mean inserting himself into a volatile political situation. The self-styled "reform" wing of the government—those undermining the monarchy to bolster foreign interests—enacted legislation to take control of the military away from the king. Meanwhile, Robert Wilcox, who had tried to remove the king in favor of the king's sister a year earlier, had been working as a surveyor during a quasi-exile in San Francisco. He still dreamed of obtaining military prominence in Hawaii, and prepared to sail back there, possibly with the encouragement of Princess Lili'uokalani. Rumors and news reports anticipated that the Hawaiian islands could descend into armed conflict, and strangers coming to Honolulu expected scrutiny. Charles Berger, the insurance agent searching for information on the *Minstrel*, was the son-in-law of one of the politicians supporting the royal family. Berger had connections throughout Honolulu, where John Cameron's checkered past would easily come to light, leading to unwanted attention.

Eschewing Hawaii, John zeroed in on a different reward posted for $100 (more than $3,000 today) in San Francisco, where John would not have to worry about political instability:

REWARD OFFERED. - Barque Wandering Minstrel, of Hongkong, 362 tons. Last heard of at Honolulu December, 1887. Reward for later information regarding this vessel; apply to A. S. MURRAY, 412 California street.

Leaving Hans behind in Jaluit, John scraped together enough money from odd jobs for the passage to San Francisco. Surrounded by massive buildings packed densely together, he traversed an elegant area of the city that represented a kind of "Wall Street of San Francisco." He visited the three-story Victorian-style building at 412 California Street, an architectural hodgepodge of garish decorative elements from nearly every European region.

John sat down with the A. S. Murray listed in the newspaper ad and unfurled a long-winded story overflowing with details—just not the truth. The good news according to John was that the *Minstrel* had not wrecked at all. The bad news according to John was that the Walker family had turned quasi pirate, selling the *Wandering Minstrel* somewhere in the South Pacific and absconding with the profit while the *Minstrel* was "wandering another name." (In the way whispers spread, a variation of the sensational tale later changed the point of illegal sale from the South Pacific to South America.) John revealed that the Walkers were not the "white sheep"—that is, moral and law-abiding people—they pretended to be, but instead were a family of fraudsters. The Walkers, he said, were comfortably "settled down in a land [with] no post office address."

With the fabrication of the story of the *Thistle*, the shipwreck at Morrell Island, and the claim about the Walkers selling the *Minstrel* and living in hiding, John and Hans aimed to sufficiently muddle the situation so that nobody would be looking for the castaways at all. They had found a way to murder the Walkers and the *Minstrel* crew from 2,700 miles away.

Chapter 14

WAIL OF LOST SOULS

Winter 1888–89

ENOUGH TIME PASSED FOR THE Walkers to surmise that John, Hans, and Mouwen had either drowned at sea or had survived but did not tell anyone where they were. Either way, "we [are] left to die," realized Charlie Walker, who had celebrated turning fifteen with a heavy heart from their "sea-girded prison." Elizabeth's silver tea set no longer seemed to be a promise of family memories to share, but an artifact of what they lost.

Ten months into their life on the island, there were no more yarns around the bonfire about comparable shipwreck adventures with happy outcomes. Nobody on the island had heard of a castaway experience that had lasted as long. The contemporaneous castaways from the *Henry James* only stayed on Palmyra Island for six weeks before their rescue. The castaways shifted between hope and faith, on the one hand, and fear and despair on the other.

The ever-changing bird population on Midway Atoll at times seemed to channel the emotional chaos that the castaways experienced. The birds overrunning Sand Island became oppressive and overwhelming. In the diary of an American visiting Midway Atoll years later, the repetitive wording of his entry projected a kind of psychological breakdown: "noise of every kind—and continuous,—the noise never ceases. The damnedest din one ever delighted in, and continuous,—day and night,—a boiler-plate factory would sound like an organ recital in this damnedest din . . . This eternal din doubled as the day darkened and with the darkness the noises became weird, mysterious, spooky like cries of pain in spiritland,—and continuous."

Many who spent time on the atoll noted the almost supernatural aura of the masses of birds and the accompanying sounds: "If human spirits are transformed and take the form of birds, I heard that first night . . . the cries of every unhappy soul that ever lived in this mysterious island. . . . How a superstitious inhabitant could have stood it, I don't know." Frederick succumbed to this line of thought. "Their dismal cries, like the wail of a lost soul, made us very dismal indeed." Another observer took the bird noises as a kind of paranormal demand: "the wail of damned souls crying their yearning to be liberated."

Edvart Olsen, who had been stranded in the *General Siegel* wreck and helped maroon Hans there, remembered the island's bird-centric meals fondly. "He would rather eat goonies' flesh than any meat obtainable in Honolulu," noted an observer who met him after he reached safety, recalling that he insisted "the eggs of the bird are so palatable that he used to drink them raw." The *Minstrel* castaways, however, grew disgusted by the monotonous diet. Still, eggs meant survival.

But food now disappeared for a combination of reasons. The August hurricane and subsequent storms had decimated nests and left hundreds of birds dead. Colder weather that followed the big storm also thinned out the bird and turtle populations. The birds seemed to become more aware of being hunted, eluding even the Walkers' resourceful dog. The castaways also believed that colder water and seasonal migrations pushed the fish farther out from shore.

With so many birds vanishing, the resulting silence disturbed the Walkers, representing the prospect of starvation. "Our life," Frederick lamented, "was one continual hunt for food."

Three castaways left Sand Island for Green Island to try to find more resources, though even crossing the lagoon became a chore. "The distance between the islands is about three miles," Frederick noted, "and to persons weak from hunger it is a long pull." In certain ways, spending time on Green Island also required extra work, as wells had to be dug deeper there, eight or nine feet underground, to reach fresh water.

The combination of drinking water—often unsanitary, even if technically potable—with little food produced what Frederick observed solemnly was "a mixture of scurvy, rotten gums, rheumatism, distended stomach, appendicitis and a few other symptoms, which ends in death." Without the birds, fewer predators existed to eat pests. At the larger encampment back on Sand Island, Charlie looked on with disgust as the drinking water brought up from their wells had to be strained of rats and insects, including cockroaches. Rats tended to be transported to islands on ships, and became more conspicuous as food sources diminished. After the *Saginaw*'s wreck on Ocean Island, that ship's supply officer observed, "The rats are more in evidence of late. At first small and timid, they are now growing larger and bolder; running about and over us in the tents during the night." In that situation, the castaways contemplated eating the rats themselves. On Sand Island, little white crabs came out at night, glowing like tiny ghosts while they gnawed on castaways' wounds. Onto those wounds would then crawl dermestid beetles, which ate dead or decayed flesh, while a variety of spiders multiplied.

As for marine life, even when mullet swam close by, fishing for them came with risk. At more than one point, a shark followed the schools of mullet into the shallower water, where the predator would sometimes turn away from the fish in favor of turtles, slicing open the shells with their teeth and then devouring them, clouding the water red with blood.

With each source of food depleted, the start of 1889 brought desperation and then catastrophe. One *Minstrel* castaway, whose name was not recorded, took out the small boat in an attempt to improve the odds of catching fish. The sea turned rough, and a wave capsized the boat. But the castaway was too weak to swim to safety. He drowned. That death epitomized the deadly combination of malnourishment, physical debilitation, and the elements. Around the same time, two of the castaways suffering from scurvy also died. One of the deceased was the crew member who earlier had begun hoarding every morsel of food he could find. Yet another castaway, Edward Dawson, was in rapid decline from scurvy.

Elizabeth's health worsened too. Her husband and sons needed to find her sustenance. As they searched, she would sit on the step into the hut. Upon their return, she would ask if there had been any luck finding something to eat. She was startingly gaunt. Frederick, also skin and bones, felt heartbroken each time he looked upon her "starved and eager" face.

He decided to take the boat to the reefs surrounding the islands, in spite of the recent drowning of their shipmate. Freddy, as the oldest of the Walker boys, could not stand letting his father go alone, even as his ma seemed to be perishing. He insisted on going out on the boat with Frederick, and they were determined to accomplish what seemed impossible. They would not come back until they had fish.

They fashioned an anchor out of a crooked scrap of iron salvaged from the wreck to help them hold a position from which to fish, though Elizabeth worried it would not be strong enough. Rowing out, Frederick and Freddy reached a patch of coral where they successfully anchored, but found no fish. Knowing they would be taking a risk, they weighed anchor and pushed out farther before anchoring again.

To their delight, fish appeared all around them, swarming the coral. A more life-affirming sight was hard to imagine. They had come full circle, in a way, to the original purpose of the *Minstrel's* voyage—though now the fishing was shorn of ambitions of profit; only survival mattered. They spotted a shark, probably a tiger shark, attracted to the fish.

While horror stories about shark attacks highlighted anthropomorphic qualities including greediness and "unreasonable pugnacity," other tales of the era mythologized sharks as guardians, gesturing toward their important role in the food chain. In one of these tales, a whaling vessel caught fire at sea and the crew ended up adrift in a lifeboat. Floating for five days, their rations running out, suffering from delirium, they glimpsed a shark. The captain supposedly harpooned the shark, which proceeded to swim through the ocean at high speed, dragging them along to safety. There was also the infamous story of the sinking *Pohoiki*, in which a Hawaiian captain and his family were taken by sharks. One of the *Pohoiki* sailors, John Manaia Nawela, clung to the shell of a lifeboat watching his shipmates die. He would later tell how he spent those hours at sea reflecting on the power of sharks as informed by his family's lore. He had a brother who had died young years earlier, and whose grave supposedly was later found empty by his mother. When she helped her fisherman husband at work in the ocean the next day, she first spotted a baby shark, which proceeded to follow her whenever she came to the water. Convinced this was her deceased son taking the form of a shark as their family *akua*, or god, she began to bring her son's favorite foods for the shark, and told Nawela on her deathbed that the shark would take care of their family. As Nawela floated among the wreckage of the *Pohoiki*, he prayed to the sharks, and would later claim that a shark took his boat on its back and carried him to shore.

For Frederick and Freddy, the giant shark in their sights could save their lives in a less magical manner—by providing food. Primal instincts came to the foreground for the Walkers. Deprived of all but the most basic gear, their makeshift spear and hooks, they had to engage in what amounted to hand-to-hand combat. To manage the bare-knuckle encounter, they wrapped their arms around the shark's head, wrestling against its powerful movements; one hard strike of its tail could sink the vessel. The fight seemed interminable before they felt the massive fish shiver, followed by a sublime stillness. The shark was theirs.

With the shark dead and secured, a rainstorm descended upon them, churning out gale-force winds. The boat moved farther from the reefs out to sea. As father and son were tossed and spun around, they lost an oar. The twisted piece of iron serving as their anchor hooked back into the sea floor, but the tighter it was pulled the more it strained. Caught in the deluge of rain in a new position, they could no longer see the island. At any moment, the rope could snap from the anchor and, if it did, they would be washed out to sea and drown.

Back on the island, with hours passing and no sign of Frederick and Freddy, Elizabeth knew something was wrong. Though deathly ill, she teetered through the storm to the beach to look for her husband and oldest son. Once night came, she would have no chance to catch sight of them. In the late afternoon, daylight running out, she spotted them in the distance amid the reefs. Charlie and Henry,

worried their ma would collapse, could not convince her to rest in
the hut, so they worked together. The three of them gathered wood
with which to build a fire so Frederick and Freddy could locate the
island.

The rain put out their flames as quickly as they could stoke them.
They needed a miracle to keep the fire burning, and they found it in
a substance derived from the natural inhabitants of the island: the
"gooney oil" made from albatross that they used for cooking. They
poured oil onto the fire whenever it was doused by the rain.

Darkness fell. They could no longer see Frederick and Freddy
on the horizon. Elizabeth, Charlie, and Henry took shifts during
the night, feeding the oil to the fire, wondering whether they could
still help their loved ones. It was the first time the family had been
split up for any significant period since boarding the *Minstrel* in
Hong Kong, and certainly since the shipwreck. The sense of a final
tragedy increased when morning brought no sign of Frederick and
Freddy. That second day passed, every minute dragging, without
change. With another night of terrifying uncertainty upon them,
Elizabeth decided that instead of the bonfire they should use a large
red lamp that had been washed onto the island from the wreck. The
lamp, filled with the gooney oil, would be better protected than a
bonfire from rain and, reminiscent of the navigation lights they
had seen from the steamship months earlier, the red glow could be
easier to spot from the reef—if their loved ones were out there at all.
Elizabeth and the boys continued to take turns tending to the lamp
on the beach all night. As hope faded with the hours, the odds grew
that Frederick and Freddy had been swallowed by the sea.

IN FACT, BY that second night, Frederick and his oldest son, in
states of complete exhaustion, were leaning over their boat clinging

to the ledge of the reef, desperately trying to prevent the little craft's anchor from being ripped away. They could not hear their family call their names, could not hear Jessie's incessant barking. But finally they spotted the light and smoke on the distant beach, which became "a beacon of hope" for them. Exposed to the storms, they suffered dehydration and intense hunger. The red glow in the distance gave Frederick and Freddy courage as they approached the third morning trapped on the reef. Frederick checked his pocket compass under the light of the lamp during the night in order to keep track of the direction of the island. At daybreak, they took their chance. With all their strength they made an "agonized effort" to swing their splintering boat around the reef, using the single oar that they still had, so they might be able to see the island.

Even after Elizabeth caught sight of her husband and son, she had to confront the fact that they might never reach land. They were struggling along the reefs, and as they moved, they were in even greater danger of being pulled to sea or smashing against the reef. She rushed over to their healthiest shipmates. She pleaded for help trying to reach the boat. Not one would come.

Time was running out, and help could come only from within the family. Charlie and Henry secured oars and swam toward the reefs. Holding out the oars, they were able to pass them to their father and brother, which allowed Frederick and Freddy to control the boat. As the two younger sons swam back, Freddy and their father steered through the reefs and finally reached the shore.

Though Frederick and Freddy's survival seemed miraculous, the three days at sea had debilitated them, and Elizabeth's constant vigil for them had set back her recovery.

All the castaways were incapacitated. At one point, George Hanker refused to come out of his kennel, and for a while Frederick, peering into it, thought the second mate was dead. He dragged him out of the structure, only to find him alive but hardly moving. He could be added to a list of those who would not survive much longer.

Elizabeth continued her decline. She attempted to remain stoic, an extension of her resolve not to complain when she was the only woman at sea. "I considered that it was my duty," a whaling captain's wife had articulated this attitude in her journal, "to bear up as cheerfully as I could." The Walkers tried to feed Elizabeth dried shark meat from the fishing expedition. The boys also dived for bêche-de-mer, or sea cucumber, which they boiled over a fire to soften, since it was easier to digest than the shark. But Elizabeth was too sick to eat any of it, and Frederick was still trying to recuperate from the recent ordeal at sea. Freddy's health, too, collapsed following their return to the island; he showed the same signs of serious scurvy that had already killed crew members. It was clear to them from Freddy's rapid deterioration that he would die without proper food.

Even in her own feeble condition, Elizabeth was horrified by Freddy's state. She knew "he was in the last stages of that dreadful disease." She fretted as she observed the worsening effects: "His legs were patches of purple, green and red. The press of a finger left a dent in the flesh." He begged for rice, which they had run out of many months before, suggesting mental confusion. Despair overtook Elizabeth.

The family now relied on its two youngest, healthiest members. Henry and Charlie had to tap into their ingenuity gained from their family's odysseys. They wandered the island searching for eggs, but they were all gone—Midway was in a state of complete desolation. They could not risk taking out the boat, which was a skeletal cross section of wooden ribs and splinters since their father and brother's fraught fishing voyage. They waded into the water as far as they could, searching futilely for fish. An idea presented itself—instead of looking for marine life in the water, they could look for any useful items on the seafloor from the *Minstrel* and *Seigel* wrecks that may never have washed ashore. To do this, they took advantage of Henry's athleticism. Searching below, there was a glimmer of an object stuck in the sand. Diving down, they discovered a bottle, which was filled with lime juice, an artifact from their ship's galley.

An airtight seal on the bottle had kept the juice fresh for what was now thirteen months of their time on the island. It was an unexpected treasure, as lime juice fought scurvy. They pried Freddy's mouth open and poured small drops down his throat, praying for him to survive another day.

Chapter 15

RIDDLES

January 1889

INSURANCE AGENT ALEXANDER S. MURRAY evaluated the tale he had been told by John Cameron about the *Wandering Minstrel,* which included the claim that the Walkers had misappropriated the schooner.

Alexander managed the San Francisco branch of the South British Fire and Marine Insurance Company, supervising their clients' cases from their offices on California Street, where they had moved on New Year's Day 1889. The new office's proximity to the post office and customs house helped Alexander's team stay informed about international happenings. (The insurance firm's headquarters was in New Zealand, from which both Alexander and his branch secretary, Charles Stringer, hailed.) Over the course of their decade in San Francisco, they earned a reputation for "upright dealing with their clients."

Reports stored in an intricate maze of wooden file drawers could be spread across Alexander's large custom-made dark-walnut desk, laying out the riddles of the *Minstrel* case. He could also bring his files back to his residence at the Cosmos Club, an exclusive social club with living spaces and banquet facilities at 317 Powell Street in Union Square.

Even though John Cameron's story may have been far-fetched, Alexander could not disprove it. But he also could not quite believe the Walkers had turned into criminals. Besides, it was John looking to gain from receiving a reward. Maritime fraud tended to inspire imitators, and John may have been galvanized to try to commit fraud by stories such as the burning of the *Brussels*, which was heavily covered in the local newspapers as well as in insurance industry publications. Fortunately, the South British Company's office had a veritable library of records, including newspapers and listings of missing and wrecked vessels. Alexander received help from his team, who sat in the office between wire partitions, including secretary Stringer and their cashier Frank Agnew.

Recent reports did not reveal any shipwreck of a vessel called the *Thistle* that John claimed to have survived, though there had been a *Thistle* wrecked near northeastern England six years earlier. And if John's story of the *Minstrel* were true, where were the original crew members? Did they remain on board as a new owner unlawfully took the vessel? By examining the nooks and crannies of shipping news over the last year and a half, the archives of the Honolulu *Daily Bulletin* revealed an interesting item.

> *Captain J. Cammeron, formerly master of the lost steamer Planter and afterwards commander of the schooner Emma, has shipped as Chief Officer of the Wandering Minstrel.*

The entry had been printed December 2, 1887, shortly before the *Minstrel* departed from Honolulu. The last name was misspelled, but there was no doubt it was about the same John Cameron. In a few words it encapsulated a much larger story, including John's stint as a captain of the crashed steamship *Planter*, which fatally wounded the twenty-year-old Florence Barbour, as well as his purchase of the mysteriously burnt schooner *Emma* to carry out his grudge against the Inter Island Steamship Company. What's more, the brief column proved that John Cameron was no mere bystander who happened upon intelligence about the *Minstrel*: he was the last-minute substitute as her first mate, a demotion for a man with a big ego. This all fit together, and pointed to John lying.

Which left a big question: if not unlawfully sold, where was the *Wandering Minstrel*?

Alexander understood that if they could track John down again to question him, they might find the *Minstrel*. Then they received a tip from Honolulu: a Hawaiian sailor who had been in Puget Sound, Washington Territory (soon to be the forty-second state of the U.S.), had seen John Cameron in Tacoma. Another person in Honolulu who knew John also contacted Alexander, reporting that John had paid a thousand dollars in cash for a stake in a hotel at Port Townsend, Washington. When John had supposedly come off the *Thistle*, he was penniless, and the reward offered by the South British Company for information about the *Minstrel*'s whereabouts was one hundred dollars. Where did he get a thousand?

A fuller picture came to light with word of John having become "engaged in the opium smuggling business."

Alexander kept close ties with the Harry N. Morse Detective Agency and Patrol System, which employed suitable operatives to chase down John Cameron. As the firm's name suggested, Morse positioned his organization not just as a detective agency but as a kind of private police force, which had also put him at odds with Isaiah Lees, the chief detective for the San Francisco police, who had received credit for closing the case of the *Brussels'* fire. Morse, fifty-four, was slender, with a prominent forehead and penetrating eyes. He was more formidable than he appeared at first glance. Morse had made an impact on the thieves and cattle rustlers who populated what was already labeled the "wild west."

He had served as sheriff of Alameda County in Northern California for nearly fifteen years. For an era depicted as having dramatic shootouts and confrontations, comparatively few actually happened, but Morse was involved in several. In 1872, a wanted bandit known as Procopio, suspected of multiple murders and described as "the most notorious desperado on the Pacific Coast," was on the run in the city of San Leandro, accompanied by a reputation for firing to kill when cornered. As a deputy and two detectives entered a saloon, Procopio, seated at a table, leapt to his feet to draw two revolvers from his belt. Morse rushed through the back entrance and grabbed Procopio's throat with one hand and held a pistol to his head with the other. "Put up your hands, Procopio," Morse said, "you're my man."

Wrapping up his career as sheriff, Harry Morse started his detective agency with offices eventually located on Montgomery Street not far from the South British insurance firm. Morse hired a team described as "detectives of established reputations for integrity, shrewdness, tenacity of purpose, and of experience." The agency served private and governmental clients. Cases usually relied more on clues than on blunt force. Black Bart, the pen name of a stagecoach robber who left behind taunting poems, had eluded authorities across Northern California for a decade. When a vagabond was arrested by the police, Morse insisted police had the wrong man. Morse was able to trace a handkerchief to a series of clues that led to Charles Boles, who confessed to being the actual Black Bart.

In what he called "the most difficult case he ever had," Morse took on government corruption related to the opium trade. Smugglers found a variety of ways to sneak opium past customs, establishing a massive illegal network spanning from China to Hawaii to California engineered to get around tariffs—in the process costing governments millions in lost revenue. One of the crown jewels of the steamship fleets, *City of Peking*, which the Walker family watched nearly anchor at the Midway Atoll, was found to have opium hidden in spare boilers by rogue crew members. Even *Oceanic*, touted by the Hawaiian government as a shining example for tourists, had $15,000 worth of opium seized from a hiding place painted over and "fastened by false bolts" between the bulkhead and water tank.

Morse worked in secret for the U.S. Treasury Department and discovered through extensive stakeouts and surveillance that an organized opium ring compromised Secret Service agents, U.S. marshals, customs agents, and even the magistrate in charge of reviewing smuggling cases. Facing so much embedded power, the investigation proceeded in fits and starts, hitting blockades until a single corrupt official with a guilty conscience admitted wrongdoing, setting off a chain of consequences for others. Morse, who had weathered allegations of misconduct from the accused, had managed to send the opium trade into a spiral, fragmenting methods and alliances.

A month after harsh rulings were handed down against corrupt officials in the opium ring cases, *Wandering Minstrel* left Honolulu with the Walker family and John Cameron, destined to wreck at Midway. Now in early 1889, Alexander Murray needed detectives to find John, and Morse was their perfect bloodhound. Morse was used to being reminded by his clients how intense insurance cases could be: "Harry, I want you to go on a very important mission for me. It may require you to be absent a long time." One of Morse's detectives, Jerome Deasy, also worked on cases for South British Insurance. He was a force to be reckoned with. Only months before, Deasy had been forced to appear in court for being too rough with witnesses.

Because he was viewed by the public as a rival to the city police department's Chief Detective Isaiah Lees, Morse acted as an alternative authority for law and order in San Francisco. Though Morse was as much a master of publicity as Lees, he ribbed the chief of detectives for craving the limelight. "It's not well to do police business with a brass band," Morse said with a laugh, ironically making the comment to a reporter. A sore point between the two investigators remained the Black Bart case, a signature victory for Morse and a blemish for Lees.

Morse could also pick apart the many flaws in Chief Detective Lees's supposed quick resolution of the case of the *Brussels* ship fire. The presence of insurance fraud was clear enough, but less certain was the scope. Morse would have reviewed with interest the details, which included the aborted plan to wreck at Midway. Faced with his own case of an apparent insurance swindle, Morse's rivalry with Lees provided extra motivation—in addition to being paid for the task by Alexander Murray of the South British Fire and Marine Insurance Company—to ensure a resolution.

The detectives, along with Alexander's unassuming secretary Charles Stringer, raced to find John Cameron.

Meanwhile, John was in the process of transporting the opium salvaged from the *General Siegel* wreck on Midway halfway across the world. With custom department inspectors tracking the movement of a major cargo of opium, every location involved—San Francisco, Port Townsend, now Tacoma—also lined up with sightings of John Cameron.

Any plan by John to pay off customs officials to overlook smuggling took a hit as a result of Morse's successful operation against complicit inspectors. Disgraced inspectors had been replaced by bureaucrats of varying efficacy, though the purge of corruption was not easy or smooth. For example, a former chief inspector at Tacoma, Erwin Gardner, was forced to resign when detectives produced evidence that he was part of the opium ring. He resurfaced in Watertown, New York, transporting more than fifteen hundred pounds of opium from Victoria in British Columbia, Canada. He was arrested, and proceeded to send a telegram from jail in cipher to a conspirator, another ex-customs inspector.

Still, it was harder for smugglers like John to hide from the newer breed of inspectors, though smugglers tried to adapt. At one point in early 1889, a woman at Port Townsend was caught with fourteen pounds of opium held at her bosom by a special "harness."

For those familiar with Tacoma, finding illegal activities concealed in the city was hardly surprising, as recorded by Rudyard Kipling after coming to the United States on the *City of Peking*. "The rude boarded pavements of the main streets rumbled under the heels of hundreds of furious men all actively engaged in hunting drinks and eligible corner-lots," he wrote. "The street itself alternated five-story business blocks of the later and more abominable forms of architecture with board shanties. Overhead the drunken telegraph, telephone, and electric-light wires tangled on the tottering posts whose butts were half-whittled through by the knife of the loafer." Amid this labyrinth, opium in tin containers had been hidden in barrels to transport them—an echo of George Hanker's hiding place in the barrel on Sand Island.

Around the same time that John hid out in Tacoma, officials made the largest seizure of smuggled opium, with a value of over $5,800, in the area's history. This may have represented only a fraction of the opium John and Hans recovered from the *Siegel* back on Midway. But for the moment, John eluded both detective extraordinaire Morse and the local police officers.

Inquiries in Tacoma generated intelligence that John headed south to Olympia, Washington Territory, where he was operating a saloon. But by the time Morse's operatives could arrive, John had moved on. The trail continued, with the hunters shadowing John into British Columbia. The detectives embedded themselves with smugglers whom they hoped knew John. Both Alexander Murray and Harry Morse had connections with police officials who joined the hunt. With John having been known as one of the best shots with a firearm in Honolulu, the investigators had to be well armed, a reminder of Morse's days as a wild west sheriff. For his part, John was tipped off about the search and managed to stay one step ahead, continually "keeping out of the way." In addition, allies spread a rumor—no doubt concocted by John himself—that there was *another* John Cameron who had made his way through Washington Territory and that investigators confused the innocent sailing captain John with some other John, never mind the positive identification by multiple people who knew "both" men.

AS WHEN HE had quickly decamped from the Marshall Islands to the United States, John now abruptly sailed back to the South Seas, accepting an opportunity to helm a trading vessel—in reality, engaged in blackbirding—for the prolific Crawford & Company. He was assigned a schooner called *Ebon*, which he described as "a stanch and handsome two-masted schooner of fifty tons." He headed for the Spanish-controlled islands of Kusaie (later called Kosrae), flying an American flag. The tactic of using a flag of a country that did not have a strong presence at an island allowed blackbirders an air of protection without subjecting themselves to actual oversight by a consulate. For one ambitious to achieve success in blackbirding, Kusaie was a place of legend as a staging ground more than a decade earlier of Bully Hayes, arguably the most infamous blackbirder. The hulking Hayes was a con artist and thief with a background in vaudeville performance. Hayes's ship had wrecked at Kusaie, where he turned himself into a veritable warlord, ruling over islanders and traders through threats of violence. When a British naval vessel came with officers aiming to arrest him, Hayes escaped in a small boat constructed from the wreckage of his ship. Several years later, Hayes returned to Kusaie in another vessel, in the process absconding with his benefactor's wife, a series of events that ended with Hayes murdered with the shore in sight.

John Cameron now followed in Hayes's blackbirding footsteps by rounding up a crew of islanders from Kusaie.

Meanwhile, a U.S. naval screw gunboat called *Alliance*, named for an earlier naval vessel from the Revolutionary War, headed to Kusaie. At the helm was Henry Clay Taylor, forty-five, trim and bright-eyed with a neatly groomed mustache and serious demeanor. Taylor's career had been shaped as an ensign during the Civil War on the steam sloop USS *Iroquois* as it hunted for the elusive Confederate ship *Shenandoah*, which intercepted vessels aiding the Union.

Part of the Asiatic Squadron, Commander Taylor's USS *Alliance* roamed the Pacific to protect American interests, particularly in regard to trade and general commerce with China and Japan. Upon reaching the harbor of Kusaie, the Stars and Stripes on John Cameron's *Ebon* attracted the captain's attention. Taylor hailed the *Ebon*. He then ordered Ensign J. C. Drake, thirty-four, an eager midwesterner, to inspect the schooner.

Ensign Drake boarded the *Ebon*. The blackbirding schooner would have been loaded with weapons. "The master of a trading vessel," John would comment, euphemizing the blackbirder, "must always be prepared to defend himself." The ship also had hiding places that could stow those weapons, including *Ebon*'s water tank. The naval ensign would have been only lightly armed, but not even Bully Hayes himself would have started a confrontation with the U.S. Navy, particularly given *Alliance*'s armaments, including rapid-fire guns of various sizes. John explained his purpose in the region by pointing to a contract he secured to recover articles from the recent wreck of a British yacht called *Nyanza*, which had run aground on a reef at Ponape (later Pohnpei) at the Caroline Islands.

Ensign Drake felt uneasy during his visit. He noticed people on the vessel exhibited what he called "a general dissatisfaction." The crew was in disorder. There was "considerable drinking."

If Commander Taylor had decided to send word to nearby islands for more information, he could have found out that *Ebon* was not registered as an American vessel, which could in turn shine a spotlight on more of John Cameron's deceptions since the time he first arrived in Mile, possibly even providing clues that John had hid critical intelligence about *Wandering Minstrel*. Reports of the naval crew's serious concerns about *Ebon* would eventually make their way to Yokohama, to the desk of the commander of the Asiatic Squadron.

But that would come after the fact, and after a stack of paper-work. For now, Commander Taylor listened to Ensign Drake's description of what he saw on the *Ebon*. Taylor decided to board John's vessel himself. But after a brief inspection, he announced he could not "justify my interference," and directed *Ebon* to proceed. John slipped away from yet another investigator.

John's pursuers had to look under every proverbial rock, but his task was simpler. Every day that he stalled meant a day closer to the death of witnesses on Sand Island—in fact, he counted on the probability that the Walkers and his other former shipmates were dead already.

Chapter 16

RACING TIME

February 1889

JOHN CAMERON'S COCONSPIRATOR, Hans Jorgensen, would have been even harder for investigators to track down. While John maneuvered through the busy ports and streets of San Francisco and Tacoma, Hans had been in and out of places altogether missing from many maps.

Hans needed capital to fund his campaign for vengeance against his former *General Siegel* shipmates who'd marooned him. He remained in the Marshall Islands to work for Crawford & Company, which engaged him to build a small schooner. If the first authorities Hans had met on Jaluit ever picked up on hints about his past that might have led them to inquire further, they'd run out of time to act. Postings tended to be short, and Consul Morgan and Imperial Commissioner Sonnenschein stepped aside for other representatives from the United States and Germany, respectively.

The schooner Hans helped build was called *Thresher*, on which he began captaining blackbirding expeditions to various islands. A thresher shark, sometimes known in that era as a thrasher shark, instead of attacking outright, could use its deadly scythe-like tail as a surprise weapon. It was an evocative name for a vessel that represented itself as offering legitimate work opportunities but that in reality hid a form of enslavement.

Like John on *Ebon*, Hans unlawfully raised an American flag over the schooner. Meanwhile, Hans gained a reputation among the profiteers for deploying brute strength while recruiting—and capturing—islanders to ship off to foreign lands. He strutted around with a heavy rifle slung on one shoulder and wore a belt holding a half dozen revolvers and a couple of knives, while a dog trained to be "ferocious" shadowed him. Hans received nicknames including the Earl of Nonouti and the Duke of Tabiteuea, references to two of the Gilbert Islands where he exploited laborers, and also enjoyed the moniker South Sea Pirate. An aura of violence surrounded Hans, who even in conversations with strangers made "no secret of the number of lives he has taken."

Rumors circulated that he killed a woman from one of the islands. Hans also married a native woman, Neiterei, whom he mistreated. While they were living on Nonouti Atoll, Neiterei asked that they take in a girl known as Sarah, whose exact age was unknown, from Makin Island, where Neiterei was from. Sarah was the daughter of a British trader and a native woman, a so-called half-caste whose status existed in limbo between communities. One morning, their visitor was on board the schooner alone with Hans. He sexually assaulted her.

Neiterei discovered this and threatened to shoot Hans, then promised to abandon him. "The very first vessel that comes in," she told him, "I will leave you." Hans swore he would beat her if she told anyone what happened. Very likely, her life was in danger, with Hans contemplating the risks to him if he let her leave. But he ultimately relented. "Wait till the vessel comes, and you can go," he said. She did just that. Meanwhile, Sarah fled under the protection of a British trader named Joseph William Henty, who vowed to protect her from Hans.

Later, a blackbirding vessel, the *Montserrat*, called at Nonouti, where Hans was gathering laborers. Upon visiting the *Montserrat*, Hans discovered his estranged wife among the captured laborers. He sought out the commander, Captain Ferguson, and told him she was his wife and that he wanted her back.

Ferguson was notorious in the region, with his crews considered "cruel as tigers . . . [who] looked upon the islanders as mere animals to be bought and sold, worked to death, and replaced by others." When Ferguson appeared, observers knew what he was after. "In the past he has been a pretty lively old seabird in his way," wrote a jaded reporter present at one of his arrivals, "and it was not thought that he had come here merely to improve his health by breathing the salt air." Ferguson had passed through the islands where Hans now resided the year prior, but his brig sank—resulting in the deaths of an estimated five hundred islanders, including a hundred children. Now that Ferguson returned, he was confronted with demands to know what happened to the islanders' loved ones. One of the islands' kings questioned Ferguson about the stories of the brig sinking: "Did the *Tahiti* turn over?" Ferguson replied, "Yes, she did capsize." Then the captain spun a lie. "But [the brig] was seen by a passing vessel that brought the news to San Francisco, and a man-of-war was sent out and we were all picked up. I got another vessel and we continued the voyage and arrived safely in Mexico."

The king looked him in the eyes. "Are you telling me the truth?"

"Why should I lie to you? Why should I be here if everyone was drowned?"

Ferguson fabricated elaborate explanations to islanders about why their loved ones had not contacted them, and promised their return as soon as their labor contracts expired. He was feted and celebrated by the relatives of islanders whose deaths he had caused. Cruelly, he told one mother who asked about a son who drowned on the *Tahiti*: "He is well and happy, and has grown taller than I am. He will be back in about two years with lots of money, and will buy you some land."

Now that he was approached by Hans about the native woman in his custody, Captain Ferguson saw an angle to exploit. Ferguson offered to trade Hans's estranged wife, but only for the right price. "If you get me some laborers," he said, "I will make the exchange, if the girl is willing." The officers of *Montserrat* aimed for a big profit. They planned to bring laborers to Guatemala to work as sugar planters. Ferguson would receive $100 a head and planned on gathering four hundred laborers. Blackbirders would use enticement and intimidation in equal measure; if a village resisted them, they would not hesitate to burn houses and destroy food.

Hans agreed to the bargain. He took the *Thresher* to other locations to find laborers. "His ferocious appearance," observed a sailor who was part of his crew at the time, "struck terror to the hearts of the natives, who ran away from him." He would march across the beach, stopping at every house to cajole islanders to come with him. When he failed to accumulate enough recruits for Captain Ferguson, he rounded up children, all under the age of ten, and placed them in the skiff, or small boat, to take them to the schooner. Parents threw themselves into the water to hinder Hans, some throwing rocks at their children's captor. Hans drew his two pistols and began firing. His bullets did not strike any of them; it was unclear whether he was trying to ward them off, or was a terrible shot. Others had noted "he could not hit the side of a barn."

Returning to *Montserrat*, even the hardened blackbirder Ferguson was appalled by Hans bringing such young children to him. Some onboard had begun to consider Hans insane. One observer wrote: "I have an idea from his random expressions and his peculiar actions at that time that he is a bit light in the upper story." Ferguson immediately ordered the children returned to their homes. As for the bargain they had struck, Hans ultimately brought twenty-two adult laborers to add to the *Montserrat* prisoners, fulfilling his deal with Ferguson, but Neiterei refused to reconcile with him. Hans insisted, even threatening legal trouble from British military tribunals. As Neiterei recounted, Hans "came on board and asked me to be his wife once more. I said 'No,' and reminded him of his outrage on [Sarah]. He said that both I and the girl lied, and that I was not to talk about it, and as it was a lie I should be made fast on board a man-of-war." Ferguson ultimately refused to help Hans further, replying that under the terms of their deal "she need not go unless she desired to."

Hans's circumstances grew precarious with the appearance of the British warship *Royalist*, an armored sloop commanded by the redoubtable Captain Edward Henry Meggs Davis (no relation to George Davis, who was captain of the U.S. Navy's *Juniata*, which had attempted to sail to Midway). A twenty-one-gun salute accompanied the raising of *Royalist*'s Union Jack. Davis handed over proclamations to Captain Ferguson and other blackbirders, as he did at other islands: "Now, therefore, I, Edward Henry Meggs Davis, Captain in Her Majesty's fleet, and Deputy Commissioner for the Western Pacific, commanding Her Majesty's ship 'Royalist,' do hereby proclaim and declare to all men

that . . . [the] islands have been placed under British protection."
The officers confiscated weapons and arms from the islanders.
Davis instructed blackbirders, whom he despised, that the native
inhabitants were subjects under British law who could, among
other restrictions, "not be deported while under the influence of
liquors or opium, or of terror produced by the use of fire-arms or
explosives."

Throughout a number of campaigns across Asia, Captain Davis,
unassuming in appearance, with delicate facial features, proved
a "very zealous, intelligent officer." Davis could be said to have
been a naval version of Frederick Walker. Also in his midforties
and hailing from Ireland, Davis was thorough but, like Frederick,
could be inflexible to the point of being self-defeating.

He was fascinated by the cultures he encountered. He collected—or
looted—hundreds of artifacts as the *Royalist* sailed along, including a
bowl made of polished coconut, an ornament made of a sperm
whale tooth, and a dagger made of a shark tooth. He also secured
a wooden coffin carved in the shape of what he thought was a
shark, with the remains of a chief inside.

Davis embraced his role not only as warship commander
but as Resident Commissioner of the islands he claimed as
protectorates—serving as investigator, prosecutor, and judge in
criminal matters. For example, he charged an American sailor
with having threatened an islander a year earlier. "He then condemned me forthwith," the shocked sailor in question wrote,
"said a trial was unnecessary." In another instance, Davis set out
to punish the murders of a European trader and several native
employees. He demanded that the suspects' village surrender
them to the *Royalist*, and when they refused, he took hostages.

Captain Davis heard rumblings about Hans Jorgensen and
opened an investigation into him. Even amid rumors and stories

spreading about him, Hans's past remained cloudy. As Robert Louis Stevenson wrote in a letter to a friend about the South Sea islands, "the Pacific is a strange place, the nineteenth century only exists there in spots." An individual could all but vanish, or completely recreate himself. Some of those who met Hans in the South Seas knew him as Jergersen, a difference of a few letters that helped further obscure his identity, while at other times he appeared to have used Jergens or Johnson, and he alternated between using Hans and Adolph (his middle name). There was also the confusion he facilitated about whether he was Danish, Swedish, or American.

Beginning an investigation that spanned multiple islands, Davis ended up in Jaluit, where Hans had gone from Mili after the ocean passage from Midway. The naval captain gathered piecemeal information. He found that at one point Hans had registered at Jaluit's Imperial Commissioner's office as Danish. From there, Davis traced Hans's activities to other islands while sailing the *Thresher* under American colors.

Davis wrote to the U.S. commercial agent on the Butaritari Atoll inquiring about any affiliation the vessel and the blackbirder might have with the United States, neither of which could be verified by the agent.

The captain called for Hans to come aboard the *Royalist*. Hans attempted to stall, but the invitation was presented to him by well-armed officers "in terms that he could not refuse." Once Hans was on board the *Royalist*, Davis asked if Hans was American, pointing out the flag on the *Thresher*. Hans claimed he was an American citizen but made his reply "with some hesitation." Davis confronted Hans with the information gleaned from Jaluit that neither Hans nor the *Thresher* was registered as American, at which point Hans equivocated, remarking that he had submitted "intention papers," documents indicating a plan to become an American citizen.

Captain Davis ordered that Hans surrender the American flag from the *Thresher*. "There are too many men of Jorgensen's stamp cruising about the world under false colors," Davis complained, "and I was glad of the opportunity of letting this man—whom I considered a disgrace to any civilized community—know that he could not shelter himself under a foreign flag with impunity." Davis later sent the flag to a U.S. consulate in Sydney, Australia.

Though Hans's blackbirding, like John Cameron's, was out in the open, authorities rarely punished such activities. Even the infamous Dr. James Murray, who had disguised himself as a missionary before massacring imprisoned laborers, later blended into polite society after agreeing to provide testimony against his shipmates, despite his own father calling him "a disgrace to creed, country and family." When questioned, Hans was garrulous, and hardly as circumspect as his former fellow castaway John Cameron. He positioned his experiences as "rivaling the adventures of Robinson Crusoe." He claimed he was stranded on (nonexistent) Morrell Island with one other castaway, presumably meaning John Cameron, erasing Mouwen from the fanciful account. Hans claimed to have eaten his own boots to survive. His fabricated version of events continued with the addition of a trading vessel that spotted the castaway duo and picked them up from Morrell, only to wreck on the same reef, after which two of the new vessel's crew members—fictionalized composites, it seemed, of the *General Siegel*'s Captain Sanders and William Brown—disappeared on the island, before the remainder were picked up by yet another schooner.

Hans could not so readily obfuscate other events, however. Captain Davis learned about Hans's sexual assault of Sarah. "I considered it my duty," Davis later reflected, "as the only authority present in this group, to inquire into the case without delay." When he was on Butaritari, Davis called for Sarah to be brought to him on the *Royalist* from Makin Island to tell her story. He also took statements from Joseph William Henty, the trader who shielded Sarah from Hans, and Neiterei, Hans's estranged wife.

Hans was tried on the *Royalist* for the rape of Sarah. Such tribunals were carried out swiftly at the impetus of the colonizers, as happened when Captain Davis tried and executed a native man who had murdered a trader. In another case led by Davis, he became executioner himself, and relished a sense of justice on behalf of the victim, a sailor named Sam Craig. "Having previously satisfied myself as to his guilt," Davis wrote of the convict in question, "from the evidence of natives who witnessed the murder, and he himself having confessed to the crime. I shot him on the same spot he murdered Craig."

During Hans's two day trial, the defendant denied everything, describing a conspiracy between his then-wife and a "mischievous" girl. But "the evidence speaks for itself," Davis insisted. After the testimony of witnesses, Captain Davis declared Hans guilty.

Knowing the likelihood of being killed, Hans was terrified. But he was helped by a technicality, or so it seemed. Davis realized Hans's crime occurred prior to the island becoming a protectorate of Great Britain. So instead of executing him, he banished Hans from the islands.

But that was not the whole story. Davis's officers loaned guns to the native policemen with their quiet blessing to execute Hans outside official military channels. This tactic fit another side of Davis's personality, his deep-seated desire for retributive justice, and also corresponded with other instances in which the officer used islanders as surrogate colonial law enforcement.

Hans was blindfolded by the islanders and his arms tied behind his back to a tree. At that juncture, as he waited to die, there finally may have been enough pressure on Hans for him to offer to reveal information about what really happened to the *Wandering Minstrel* in return for clemency. Before that could happen, the *Montserrat* returned to the island and Captain Ferguson, having heard what was happening, rescued Hans. However little personal affection Ferguson had for Hans, he was not interested in setting a precedent in which islanders punished foreigners. Ferguson's livelihood, after all, comprised a series of violent acts against such islanders.

WITH THE LONG line of investigators—insurance executives, private detectives, police officers, and sea commanders—no closer to answers about the *Wandering Minstrel* from the only people who had them, John and Hans, the loved ones of the Walkers and other crew members were left to fear the worst.

Most vocal among those loved ones was Elizabeth Walker's brother, J. L. Eyton, forty-four, who lived with his wife, Kiyo, and their four children in the busy port city of Yokohama. They were close to Elizabeth and Frederick and the children, who had been brought up in that city. J. L. and Kiyo awaited news of the *Minstrel*, but as time passed they became despondent.

Yokohama stood out as one of Japan's commercial centers for trade and shipping, showing off significant investment in the piers and landing places. More than three thousand non-Japanese residents lived there. J. L., after sailing on a British vessel to Yokohama in 1867, had become involved in a brewery business before working as reporter and manager of the *Japan Gazette*, one of the five newspapers in Yokohama published for English-speaking readers. He and a colleague, Barbin Pratt, left their newspaper jobs to start an auction house that sold possessions on behalf of estates and dissolving businesses.

J. L.'s doggedness in solving problems was regularly tested by the drama that auctions attracted, sometimes turning him into a peacemaker or even impromptu investigator. While J. L. conducted one auction, an alarm was raised when the watch guards, in which pocket watches were kept, of two of the men in attendance were spotted dangling empty. J. L. and the auction-eers immediately sealed the doors of the auction rooms, leaving everyone a suspect. J. L. helped ensure that each individual was searched, but nothing was found, with the "dexterous thief," as a local newspaper surmised, "having evidently escaped before the alarm was raised." The thief had a good eye for watches. One of the victims, Friedrich Retz, was a well-known watchmaker and importer whose stolen gold watch was worth twenty-five pounds.

Family deaths and other losses, the province of auction houses, tended to bring out vulnerabilities and high emotions. Weeks after the pickpocket stole the watches at their auction, J. L. and his team arranged an auction of all the furniture in a mansion, when "curiously enough" a fire broke out and burned down the home, insured for $8,000, and destroyed all the furniture, in-sured for $2,000.

J. L.'s firm also handled auctioning the remains of shipwrecks. They would divide up the rights to the cargo and to the ship itself, sometimes spurring bidding wars. How heartbreaking to think of the same being done for his sister's schooner if it were one day discovered and the Walkers' deaths confirmed. The un-certainty weighed on J. L. and Kiyo. A woman whose brother was on another ship that vanished at sea would mark his birthday every year with the kind of anguish that J. L. and Kiyo began to feel. "Oh, that we knew his fate," the woman wrote on one of these birthdays. "Of a whole vessel's crew, probably not one was left to tell the sad tale."

By this time, Kiyo, thirty-nine, fell ill. In great pain and not expected to survive, she pined for her beloved sister-in-law, Elizabeth. With each passing day, more people resigned themselves to a sad assumption: the Walkers perished with their ship. But Kiyo would not give up. She told J. L. she wanted to consult a soothsayer, or fortune-teller.

Japan had a rich tradition of spiritualism. Some fortune-tellers stood on street corners to attract customers, holding tubes with thin sticks attached to numbers that corresponded to predictions about that future that were held in a small set of drawers. Others worked with "magic" mirrors framed in bronze to reveal hidden truths in an individual's reflection. One foreign reporter watched these "wizards" with a skeptical eye as they peddled their services. "There are fools even in Japan, so our fortune-teller drives a pretty fair trade, on the sure capital of human credulity and personal impudence."

J. L. viewed himself as a rationalist but also believed the sum total of an individual's mental capacity included a soul with preternatural qualities. He explored this notion approximately six months earlier while serving as a member of a coroner's jury in the case of the death of a man suffering from mental confusion, who had been convicted of arson. "That combination of the whole brain which makes the soul," was the formulation generated by a discussion between J. L. and a physician in those proceedings. At the very least, Kiyo's desire to grapple with spiritual mysteries and her desire to meet with a soothsayer might put her at ease.

The fortune-tellers selling their services to the public in the streets of Yokohama tended to be men. Kiyo had reason to be more intrigued by the mediums known as *itako*, blind women whose practices connected to the mystical qualities believed to exist around Mount Iwaki, in the Aomori Prefecture, the same section of Japan as Yokohama. When the *itako* were young girls, as one scholar described it, they "had to undergo for two to five years severe training involving daily cold water austerities, chanting practices, and the like, until she was 'ready' for the initiation rite in which she was ritually wed to a deity." The *itako* were believed to facilitate communication between the living and the dead, which could in turn allow healing and prophecy. They sang as part of their communications with the spirits, connecting with the musical education of Elizabeth and J. L.'s childhoods. According to believers, the spirit world's inhabitants could send warnings through these human conduits about events that they did not want to come to pass.

At times, the government, feeling threatened by the influence of mediums, criminalized them. Aomori Prefecture issued an edict in 1874 that applied to *itako*: "It is hereby decreed that when such persons are discovered, they will immediately be arrested and punished." The community persisted in consulting them, and the *itako* risked their freedom to share what they believed were unique powers.

The soothsayer Kiyo consulted told her she had a vision: "[Elizabeth Walker is] still living, but in great distress, and longed to get away from the place they were in." This renewed Kiyo's faith.

Kiyo begged J. L. to do something. But unless a soothsayer could summon a nautical map of the castaways' location, and thousands of dollars, they could not raise a search party. Still,

J. L. knew many people from the auction house and the newspaper, and he could manage to determine which vessels prepared to ship out and where they headed. From the waterfront, as described by a contemporary guidebook, "whole fleets of junks and fishing sampans may be observed as they sail away or are propelled by the labor of the muscular crew." J. L. could confer with merchant captains and sailors and urge them to look for any sign of the *Minstrel*. Captain Seabury's *City of Peking* arrived from San Francisco. In addition, a couple of other massive steamships, *City of Sydney* and *Oceanic*, that had started in Hong Kong, stopped in Yokohama on the way to California. But none of those ships' captains could agree to deviate from strict schedules to look for the *Minstrel*.

Another possibility presented itself, however: Charles Johnson. He had command of a new white-planked schooner, the *Norma*. The schooner had launched four months earlier from the Ishikawa shipyards after being built to order with seventy tons of cargo storage, a relatively modest amount, at sixty-one feet long with a seven foot draft (the distance from the waterline to the bottom of the hull). The vessel was constructed from a variety of Japanese woods, including frames of keyaki hardwood, a deck of hinoki cypress, and outdoor planking of sugi cedar. Charles could manage voyages with as few as six men, including three seamen, a steward, and a mate.

As luck would have it, the *Norma* was setting off on a sharking voyage. J. L. knew this meant the schooner could visit some of the same spots known for sharks where the *Wandering Minstrel* dropped anchor. Despite the absurdly long odds of asking sailors to look for a lost vessel and for his sister's family in the largest, deepest ocean in the world, J. L. sought help from any quarter.

Charles Johnson, however, was not the most approachable person in Yokohama. Few knew much about him other than the fact that he was an American from Massachusetts. Sometimes, too, he went by the name Charles Smith. They said he smuggled opium into Hawaii and the United States.

Charles, like many other smugglers, conceived of opium as a stepping-stone to other ventures. As observed by another smuggler connected to the *Norma*, "many of the rich and prominent men . . . had laid the foundations of their fortunes through [opium]." Illegal trafficking came with the need for secrecy, expediency, and sometimes violence. Charles had to think steps ahead. Transportation of contraband also required having hiding places throughout the *Norma* that would be missed by inspectors, turning the vessel into "an opium smuggler of the first magnitude."

When J. L. had the chance to speak to Charles, he tried to overlook the sordidness of the captain's career. Charles was an athletic man in his early forties with an air of confidence that verged on arrogance. He had an honest face but a furtive manner, glancing around at the slightest movement or noise around him. His distinctive feature was a lack of eyelashes, which could have resulted from any number of conditions. He proved an easy conversationalist despite the jittery mannerisms. J. L. explained the situation of the *Minstrel*, which had also been publicized in newspaper articles. He pleaded that Charles make "particular examination" of the islands on his way through the Pacific and keep a "sharp lookout" for any sign of the *Minstrel*'s fate.

Charles thought the whole idea was foolhardy, a "forlorn hope." But he wasn't entirely dismissive, knowing an angle for making money could come from unexpected quarters, including the rewards offered for the *Minstrel*. He had lost every penny he had not too long ago. But he held on to the quintessentially American certainty that setbacks were just bad luck, while good luck was destiny. Maybe at the next port, on the next voyage, his fortunes would turn around.

Chapter 17

LETTER IN A BOTTLE

February 1889

THE WALKER FAMILY KNEW THEY were dying. Despite the gradual return of birds to the atoll, the castaways could barely muster the strength to hunt. All of them, as Elizabeth expressed it, were "in a very low state." Even she ran out of hope: "A miserable death seemed staring [us] in the face." As another woman of the era sailing with her family articulated, with every voyage at sea they felt "a sad heart, knowing not whether we should ever behold the faces of friends near and dear to us again on earth." This was truer than ever. Elizabeth could not help but second-guess herself. If she had allowed Frederick to kill Hans Jorgensen, even though it may have led to further violence with John Cameron, they at least might have had a chance to take the reconstructed boat or to send trusted shipmates.

Elizabeth and the boys prepared themselves for Frederick's imminent death. "He had become so weakened and despondent that we quite expected to lose him, for we had no strength-giving food," Elizabeth chronicled. "He could not eat the seabird goonie, and it was very seldom that we could procure fish for him. Occasionally we did discover a few crows' eggs, and these I beat up in water for him, but each day saw him weaker and weaker." Freddy was arguably doing as poorly. The crew lost all will to help each other, though the Walkers still felt responsibility toward them. Those who had enough stamina continued to fish when possible, "but they refused to contribute in the slightest degree to sustain the lives of the unfortunate family of their captain." Among the crew members, Edward Dawson, like Freddy, continued suffering from extreme scurvy, compounded by overall physical deprivation.

On March 17 around five o'clock in the evening, Frederick dragged himself up to chop wood before dark. They ran low on kindling and needed more so that smoke from a fire provided a signal in the faint chance that a ship came near. The once-routine task of heaving the makeshift axe took a physical and psychological toll on Frederick. He knew in his "deplorable condition" the smallest exertion could be his last.

As he chopped, he could not believe his eyes. Maybe he was hallucinating. He saw something in the distance. A shape like a sail. He was flooded with "unexpressible feelings." The supply officer of the *Saginaw*, George Read, had once described a similar moment on Ocean Island: "Far off to the northeast and close to the horizon there was something like a shadow that had not been there." In that moment, counterintuitively, fears increased. "She was not heading directly for us," Read wrote, "and I cannot describe the anxiety with which I watched to see if she was going to pass by, my heart was thumping so that one could hear it."

Experiencing the same mixture of amazement and panic, Frederick spread the word. George Hanker crawled out of his kennel in a rush of emotions "better imagined," as he put it, "than described."

As they had learned from the sighting of *City of Peking* four months earlier, they could not allow themselves to take for granted that a vessel would anchor. They had to overcome their physical frailties and act quickly. The firewood on which Frederick used every ounce of his energy to chop became critical to survival. Elizabeth took charge: "We carried wood to the tops of the highest hills and made large beacon fires." They also put up a flag.

THE SAIL THEY had spotted belonged to Charles Johnson's *Norma*, though Midway had not been his initial destination. After leaving Japan in early February 1889 following his conversation with J. L. Eyton, he charted a route five hundred nautical miles southeast of Yokohama to the Bonin Islands, a remote series of more than thirty volcanic islands. "Bonin" was a rough transliteration of the Japanese word for "uninhabited," which proved not quite accurate. One of the islands, named Peel, had attracted a "motley colony" from other locations to form a settlement. But the other islands in the cluster had indeed remained uninhabited other than for relatively brief sojourns by castaways or explorers, who had gestured toward claiming the islands at various times in the name of Great Britain, the United States, and Russia, before Japan officially claimed them.

The Bonin archipelago, unlike the many remote, desolate islands in the Pacific, overflowed with tropical vegetation, shady forests, ample fish and massive turtles, wild animals including deer, fresh water from running streams, and rocky cliffs and caves for shelter. Bonin came about as close as reality could to the lush paradises depicted in fictional "stranded" narratives, including *The Swiss Family Robinson*. Some castaways over the years voluntarily remained after their shipwrecks.

Almost exactly a year earlier, Charles found himself on the Bonin Islands during the defining crisis in his life. At that time, he had reached the peak of a long obsession with a story of a lost treasure, filled with betrayals and twists that could rival Robert Louis Stevenson's *Treasure Island*.

That story of the treasure that so captivated Charles went back more than sixty years, to the revolution in Peru for independence from Spanish control. Realizing their wealth was in danger, a group of Spanish royalists in Lima chartered a brig, the *Peruvian*, which they piled high with treasures and property worth many millions of dollars.

A Scottish soldier of fortune named Andrew Robertson heard about the vessel loaded with treasure. He wanted to make an impression on a married woman he coveted, so in the middle of the night, he and a dozen mercenaries overpowered the *Peruvian*'s crew and sailed away with the treasure-laden vessel. Accounts indicated that Robertson eventually buried the majority of the treasure on Agrigan (or Agrihan), a volcanic island of the Mariana archipelago. Robertson and his remaining accomplices found passage to Honolulu, where they claimed to be survivors of a shipwreck.

A sea captain found out the truth and placed Robertson in chains on a fishing vessel, demanding that he lead him to the treasure. Instead, Robertson jumped overboard and drowned himself, leaving the treasure lost.

Decades later, Charles Johnson aimed to change that, collecting enough maps, charts, and documents on the topic to form a small library. He even tracked down elderly eyewitnesses to Robertson's escapades. Charles felt certain he had among his materials "the clue to the secret of this wealth." By late 1887, Charles recruited a former schoolteacher, James Forbes, as a first mate, and hired a crew. In Yokohama, they boarded the *Nereid*, a nimble craft forty-six feet in length and sailing under the American flag, for a voyage to recover the treasure of Charles's dream turned obsession.

Reaching the southernmost Mariana Islands, Charles went

onshore, leaving Forbes in charge. But the first mate wrangled most of the crew, by force or intimidation, into a scheme to take the schooner. The former schoolteacher began "pulling down the stars and stripes" and replaced it with a black umbrella cover that he raised as a pirate flag as they "shot out of the harbor."

When Charles returned, the *Nereid* was gone. Once word reached the U.S. Consul in Japan, an American warship called *Monocacy* "started in pursuit of the pirate" but came up empty-handed. From what American consuls learned through diplomatic correspondence, Forbes, branded the "American Captain Kidd," sold the *Nereid* for six hundred dollars and "as soon as he got the money, Forbes disappeared and has not been heard of since."

(The international press followed *Nereid*'s escapades closely in 1888, in time for the actors in another unfolding sea drama to read about them when they arrived in Jaluit—John Cameron and Hans Jorgensen. Learning about the *Nereid* being sold out from under its rightful owner may well have inspired John's fabrication to the insurance agents about the Walkers doing the same, and John and Hans could have imitated the lie told by the original pirate thieves of the *Peruvian* who claimed at Honolulu to be survivors of a fictional shipwreck.)

Meanwhile, a distraught Charles Johnson secured passage to the Bonin Islands, where he waited for roughly a month, steeped in bitterness and regret, until a steamship that passed brought him back to Japan. Charles was left "all but destitute." His losses included the estimated $3,500 value of the nearly new *Nereid*, and the incalculable loss of the treasure itself, had he succeeded in finding it, "a project which [had] now been destroyed" with the disappearance of his research materials. None of this accounted for the debts he accrued searching for the treasure, nor could there ever be an official accounting of opium hidden on the *Nereid*.

Now a revitalized Charles was back on picturesque Bonin to redeem his career on his newly minted *Norma*. On February 4, 1889, he ordered the *Norma* to depart from Bonin, finding "fresh breezes" waiting for them at sea. As Charles was fond of describing sailing under such conditions, they "fairly skipped the seas." They next sailed to Ocean Island, approaching on February 26. "Strong northwest winds," Charles recorded in his log. To get to the island, they had to access the lagoon. He continued documenting his voyage: "Very high breakers; could not enter the lagoon." After three days, Charles finally navigated the schooner to a safe anchorage.

He found the remnants of the survival shed built by the Hawaiian government when they had sought castaways from the *Dunnottar Castle*. It was buried in the sand. He dug out the shed and repositioned the tanks to catch rainwater for any future visitors or castaways. He also planted the seeds he had taken with him from Bonin Island. Charles planting seeds was an example of an ecological phenomenon often carried out by sailors; as they disembarked at a new island, sailors would leave behind plants and animals from another island, sometimes forever changing the landscapes—in some of the most remote spots on earth.

Charles and crew also dug a well for fresh water. He spotted part of the wreckage still remaining from the *Dunnottar Castle* from two years and eight months prior. They used this for firewood.

They remained anchored in the lagoon for approximately two weeks before moving to the other side of the reef to fish for sharks. While opium smuggling represented the more profitable pursuit, it had become increasingly fraught. For one thing, Charles had become too well known for his own good by customs inspectors at the ports. Further, increased tensions between officials and heavily armed smugglers led to violence. Weeks before *Norma* departed from Yokohama, a native Hawaiian named Hakuelo slept at his house in Honolulu before he was dragged out of bed and beaten with steel knuckles that opened a gash in his head. Hakuelo was a customs inspector tasked with confiscating opium, and the culprits were smugglers.

All the while, crusading officials in the United States deployed "revenue cutters," sleek ships to patrol bays and harbors in order to board vessels, inspect the cargo, and search for opium. Clearly, Charles had good reason to turn his hopes to shark fishing as a way to fill his coffers. But a day after beginning to fish, on March 14, *Norma* lost an anchor. Charles tried to remain safely in place, but after two more days he decided the best course of action would be to sail on. In thinking about other spots considered shark infested, he remembered the rarely visited Midway.

He also had the voice of J. L. Eyton in his head, pleading with him to look everywhere he could for any sign of the lost ship of his sister and brother-in-law. Although he doubted that anyone aboard *Wandering Minstrel* could be alive so long after its disappearance, Charles navigated through the reef and to Green Island, which seemed deserted, before deciding he might as well continue to Midway's other island.

NOW THAT THEY had seen a sail, the castaways on Sand Island reveled in their burst of hope. Still, when night fell, they could not tell which direction the schooner moved, or find any hint whether their signals were seen, or even if the vessel was still there. They had to spend the night ensuring the fires burned. Elizabeth could hardly wait for daylight, when their final hope would either be extinguished or confirmed. Finally, "the anxiously awaited morning dawned." To their elation, with daylight came a view of the *Norma* in the harbor. As the *Saginaw*'s Read had put it, "the long dreary suspense was over; our relief was near, and I slid down . . . pale and speechless. The few moments of tense watchfulness had seemed to me like hours of suspense, and it is slight wonder that it took some time to recover my speech." The Walkers and the other *Minstrel* castaways rejoiced. From the schooner, there was the sound of a gun firing, which produced "reverberation [that] was like distant thunder."

Celebrations broke out; even in their weakened states, nothing could hold back hollering and cheering. The *Norma* was close enough to be reached by their last lifeboat, though the craft leaked like a sieve since the damage sustained when Frederick and Freddy got swept away while fishing. Still, at least it could fit six, so three of the Walkers and three crew members climbed aboard. Those who reached the schooner first would be able to eat first. Edward Dawson was in the worst shape of the sailors, and would have to be carried to the boat. From the family, Charlie had sufficient size and strength to help Frederick and Freddy, who were both failing rapidly, and to help transfer Freddy from the lifeboat onto *Norma*.

AFTER FOURTEEN MONTHS, it was hard to believe the schooner was even real, but soon the six of them boarded the *Norma* and met Captain Charles Johnson. He could see they had been reduced to skeletons. He provided breakfast. Frederick and Charlie then returned to shore, bringing with them, as a joyful Elizabeth recorded, "bread, butter, tea, sugar, tinned milk, a picul [portion] of potatoes and a picul of rice." It was such a bounty that Elizabeth and the other emaciated, ailing castaways could hardly take it all in. "We ate so much of these provisions that our weakened stomachs refused food . . . and our feast made us ill."

The American schooner captain with his conspicuously missing

eyelashes must have seemed almost a mystical savior. Amid the out-
pouring of gratitude, they could not have foreseen yet another crisis:
Charles told them that in order to take them off the island, they
would have to pay what the Walkers characterized as "an exorbitant
amount" of money.

Robert Louis Stevenson and Lloyd Osbourne fictionalized this
exchange in *The Wrecker*. Captain Trent, an older and cantanker-
ous stand-in for Charles Johnson, described to a group of castaways
what he saw that made him stop for them on Midway. "Lucky for
you I saw your signal. I didn't know I was so near this beastly island, there must be a drift to the south'ard here; and when I came
on deck this morning at eight bells, I thought it was a ship afire."

The character of Trent talks about bringing the castaways to
safety. "I'm glad of course to be able to give you a passage to 'Frisco;
one sailor-man should help another, that's my motto. But when you
want a thing in this world, you generally always have to pay for it."
He gives a "brief, joyless laugh," then continues: "I have no idea of
losing by my kindness." When the castaways speak about paying
anything in reason, Trent bristles. "Reason's between two people,
and there's only one here. I'm the judge; I'm reason. . . . If you want
passage in my ship, you have to pay my price."

Similarly, Charles claimed that by stopping at the islands while
fulfilling his promise to search for the Walkers, he "had lost his
season" for fishing and commerce. To divert his route to Honolulu,
the closest port to secure proper medical care for the castaways, "he
would have to be paid sufficient[ly] to compensate him for his loss
of time and money."

The Walkers of course did not have money with them. Even if
they promised to pay at a later date once they were safe, Frederick
and Elizabeth foresaw "practical ruin" if they accepted Charles's
terms, requiring obtaining large loans. But the other captain

remained "obdurate." From Charles's perspective, he had recently been left penniless by misfortune, the theft and pirating of his *Nereid*; why shouldn't another captain embrace similar adversity as the price of being saved? All the better that such a development helped Charles's financial recovery.

Charles also knew that the Eytons in Yokohama, so committed to finding their relatives, might pay a portion of the money.

Larger opportunity emerged too. As Charles put it, "opium was very dear in Honolulu," fetching higher prices the harder it was to smuggle past crusading inspectors. He could think back to an account told by William Seabury, the captain of *City of Peking*, who was in Yokohama at the same time Charles had prepared to sail. Here was Seabury's story: Upon *Peking*'s recent arrival at Honolulu, an argument ensued between the Hawaiian quarantine officer, who received information that two of the Chinese crew members had typhus fever, and the deputy surveyor, who suspected that "a vast quantity of opium" may have been hidden in the steerage. The two inspectors waved around statutes as each insisted on his respective authority. While no opium was apparently discovered, a temporary quarantine was initiated, which had unintended diplomatic ripples, as one of the passengers quarantined possessed important documents for the Korean delegation at Washington, D.C., which if delayed could lead to war between China and Korea. (Importantly, Charles could also have heard Captain Seabury's description of nearly landing at Midway Atoll, and his musings that one day he might find a shipwreck there.)

Seabury's account provided an epiphany. If Charles showed up in the port of Honolulu with rescued castaways in dire medical need—and not just any castaways but the *Minstrel* castaways, who were believed dead—he could use the distraction to smuggle opium right past inspectors. Charles could profit from multiple sources.

Among the surviving crew of the *Minstrel*, George Hanker was the most likely to know about Charles, or possibly to have met him before, just as prior to the *Minstrel* he had known John Cameron, who also had been part of the Pacific's overlapping rings of opium smugglers. While holed up in his kennel, George had moved in and out of states of rationality, but the excitement of the rescue brought renewed clarity, as had the earlier attempted murder against the Walkers by Hans Jorgensen. It was likely that George and other *Minstrel* sailors had heard about Charles's obsessive quest for Spanish treasure, and his tendency to keep his ships well-armed. The gun shot from the *Norma* when morning broke was not just a signal of arrival, heralding rescue, but also a warning that the castaways better meet Charles's demands.

A deadlock between the sides could easily turn violent. Some of the castaways could hardly walk, but others, reinvigorated by the prospect of a way off the island and by the food from the *Norma*, could rise against the schooner's captain and small crew. They might have been trying to embrace hope and faith, but it would become difficult to avoid the kind of kill-or-be-killed mindsets that John Cameron and Hans Jorgensen had adopted.

To tangle matters further, Frederick began to recover strength with nourishment, but Freddy did not, and once Freddy and Edward Dawson were on board the *Norma*, they remained too weak to return to shore. When Charles made his financial demands, the two dying young men became pawns in a chess match.

Stormy weather brought Freddy and Edward even farther away from the rest of the castaways. A gale threatened, risking damage to *Norma*. That led Charles to move *Norma* to the lee side of the island, the side sheltered from the wind, which was a sound sailing decision. But once *Norma* reached the new spot, the reef ripped away an anchor—the second anchor lost by *Norma* since leaving

Japan. This was an ominous echo of how the *Minstrel* wrecked more than a year earlier. Their rescue vessel would become another shipwreck if they did not resolve their differences and leave soon.

The castaways also had to worry that Charles could sail away with the two sickest of their group, either out of frustration or in a show of dominance.

As Elizabeth had managed the rising threat from John Cameron and Hans Jorgensen, she again had to guide Frederick into giving their adversary what he wanted while eliminating the danger to their family and shipmates. They developed a compromise: they would agree to pay Charles a dollar amount to be determined by the authorities in Honolulu. Treasure-seeking Charles could envision a fortune, while the authorities would employ practical accounting to consider what funds were available to pay him. Both parties agreed to the new terms, also consenting rather bizarrely to publicly characterize the arrangement as the Walkers "chartering" the *Norma*, possibly a first in the history of castaway rescues.

At the hut on March 25, Frederick wrote out a letter for any future visitors to Midway. Inside the small shelter, Frederick observed the chilling juxtaposition of inscriptions by *General Siegel* crew member Will Brown and his murderer, Hans Jorgensen, reflecting on how close the Walkers, too, came to being casualties of their enemies' machinations. Frederick jotted down in his letter advice about birds and fish on Midway, and how to find enough to eat during different seasons of the year. He placed the letter in a bottle with a cork, possibly the same bottle that had contained lime juice, and hung it up in the hut. The text survives. He included information about the *Norma* and closed the letter this way: "We intend to leave tomorrow for Honolulu, I having chartered the vessel for that purpose." He signed: "F. D. Walker, Master, Late *Wandering Minstrel*," the "late" being a farewell to the vessel that had brought them there. The information about Charles and the *Norma* may well have been included in part to leave a trail, considering Charles's volatility, in case anything strange happened on the passage they were about to make.

In the morning, Elizabeth and Frederick looked with stunned expressions over that landscape, where the birds flocked in great numbers again as the Walkers walked across the sand for the last time as castaways. Frederick later wrote: "We bid good-bye to the scene of our fourteen months' captivity."

Chapter 18

WRECKAGES AND RENEWALS

AS THE *NORMA* SET SAIL with the rescued castaways, Elizabeth and Frederick kept vigil over Freddy. Everyone worried he would not survive the voyage. Nor did it augur well for Freddy when crewmate Edward Dawson, whose scurvy case was among the worst, died two days after their departure, leaving eleven *Minstrel* crew members and the five Walkers. Captain Charles Johnson documented the loss: "We buried the deceased in the usual custom of sea burials." The *Norma*'s flag flew at half-mast. A journal from a woman sailing in that era described the rest of the tradition: "The body was laid out with great decency, and then enveloped in a hammock and sewed up. About fifty pounds of stones were secured to the feet . . . The corpse was extended upon a plank on the rail at the gangway, and all hands were called around, on the right side of the ship, to witness the solemn scene, and to join in the prayers to be offered up." Even hardened sailors teared up, then a gun was fired from the bow of the schooner. They tossed Dawson's body overboard.

With clear weather, the passage took eleven days. As they neared Honolulu on the afternoon of April 6, 1889, the *Norma* was misidentified from port as a schooner called *Triumph*. As it happened, a cross section of people, including reporters, eagerly awaited the

Triumph to come from Victoria, British Columbia, with cable wire to be used for the new enterprise to provide communication between the islands. The important potential development for commerce and government was monitored by officials there and abroad, including in Washington by Secretary of State Thomas Bayard. But none of them knew that the *Triumph* had taken a detour to go sealing when they'd discovered that the cable was not ready to be picked up at Victoria.

Interest and astonishment spread through Honolulu upon the discovery that the arriving vessel was not *Triumph* but rather *Norma*, carrying the survivors of the *Wandering Minstrel*. A reporter from the Honolulu *Advertiser* boarded the *Norma* the next morning to see the former castaways. Frederick Wundenberg, the Honolulu postmaster, sent fruit aboard. The passengers in the worst shape were rushed onto shore for medical treatment. All the commotion provided Charles Johnson what he needed in order to move his hidden supply of opium onto land. Any attempt to slow down the much-needed treatment of rescued castaways would have caused more outrage than any official could withstand.

During the passage to Honolulu, Freddy regained color in his face, a fact attributed to better nourishment and ocean breezes. But his health remained fragile, and he required immediate care.

The Walkers had other ongoing concerns upon arriving at Honolulu. The reporter told Frederick what he had heard about John Cameron's activities in Washington Territory, and they exchanged information about Hans Jorgensen, whom the reporter called a "desperado." With new intelligence gathered, the Walkers understood that both John and Hans continued to pose dangers to them. Both had conspired to let them die on Midway, and nobody knew exactly where the two men were now. Hans had committed murders on Midway in part to stop witnesses from revealing his threats to the German consul Glade, and he could try to harm the Walkers before they relayed everything else that had transpired.

Elsewhere, John Cameron's shock can hardly be imagined upon hearing the news that the Walkers and the other *Wandering Minstrel* survivors had been rescued against the odds and were in Honolulu. With typical shamelessness, John deployed subterfuge and sabotage to insulate himself. He planted a story with Captain Samuel Brooks of the *Arizona*, probably at port in Queenstown, New Zealand, but possibly on the open sea when the *Arizona* came across *Ebon*. It was a savvy ploy by John. The *Arizona* was on the way to Liverpool, and John knew that Brooks, a well-known source for shipping news, would retell John's story. The British newspapers were further removed from events covered by Honolulu reporters than the American press, and soon newspapers across England picked up the John Cameron–crafted version of the story courtesy of the *Arizona*, a narrative which in turn crept into American papers. Writing as though the *Minstrel* were still missing, these accounts noted: "Nothing was heard of her or of the crew, until the first mate and a castaway reached Honolulu, in a sailboat, on the 21st of April, when the mate gave information that the *Wandering Minstrel* encountered a fearful gale about two years ago, and foundered, but all hands reached the islands in safety." In addition to the most obvious fabrication that John (the referenced "first mate") had alerted anyone to the castaways and their location, the change in the timeline was strategic. John, Hans, and Mouwen actually reached land (Mili, not Honolulu) on November 25, 1888, not April 21, 1889. To the uninformed, the falsified later date made the next claim more plausible: "The authorities immediately chartered a vessel to proceed to the aid of the captain, his wife, children, and the surviving members of the crew, who will, no doubt, be rescued in due time." This would have come as a surprise to the Walkers and Captain Charles Johnson of the *Norma*. The Walkers and other survivors had reached Honolulu before these anonymous "tidings" began circulating, but most people who saw the account would not know that, shrouding the actual chronology of events while rendering that unnamed first mate a hero.

In a last twist of the knife by John, his earlier anonymous counterfeit reports—the accusations that the Walkers had been part of a pirating scheme—essentially forced the Walkers into a trial even as they struggled to regain their health. The British consulate at Honolulu assembled a Naval Court "to investigate the circumstances attending the wreck of the British merchant ship 'Wandering Minstrel.'" Reviewing evidence, they concluded that Frederick "managed his vessel in a seamanlike and proper manner, and, when a casualty was inevitable, to have done everything in his power to avert it." Of those who did not survive, they found "that all possible steps appear to have been taken by the master for the welfare of his crew after the wreck; that no blame attaches to the master for the deaths of any of them." The three-person panel also condemned Frank Lord's theft of a valuable lifeboat, and confirmed that John and Hans never told authorities about their shipmates after reaching safety. Interestingly, the court's report also called attention to the Walkers' "humane action" toward the marooned Hans, implicitly critiquing the previous decision by Hawaiian authorities to let the outlaw starve and die on Midway. Later reflections constructed from records of the U.S. Navy highlighted how the narrative showed "Mrs. Walker's courage standing out prominently."

No trace was ever found of Frank Lord or the boat he stole.

While the Walkers organized their affairs, second mate George Hanker spent time in and out of Honolulu hospital rooms for stomach issues, and possibly for ongoing mental struggles from their time on the island. George would sail again later on, this time on whaling expeditions, but inevitably squandered the money he earned, at which point he would return to Hawaii to find work. He had several accidents during the course of odd jobs clearing land. When recuperating, he entertained hospital staff and visitors with yarns, including those about his time as a castaway on Midway, recalling how the refugees from the *Minstrel* looked like a "hungry band of Crusoes." In addition to his avocation as raconteur, George proved to have a talent for painting and drawing, and even sold some of his artwork.

Reporters in Honolulu sought out survivors, including George, to talk about the *Wandering Minstrel* shipwreck. By far the most prominent writer collecting details was Robert Louis Stevenson. The Scottish novelist best known for *Treasure Island* and *Strange Case of Dr. Jekyll and Mr. Hyde* had sailed from California on a tour of Hawaii and other Pacific islands in a quest to renew his health as well as to gather inspiration for several new projects, both fiction and nonfiction. At five ten and 120 pounds, thirty-eight-year-old Stevenson could be mistaken for a castaway himself, looking like a "walking skeleton." He adopted the persona of a European traveler in the Pacific, easy to spot in bare feet with "thin calico shirt and a light pair of flannel trousers, with a little white yachting cap."

Stevenson thought Honolulu too "civilized" compared to outer islands of the Pacific. Still, he enjoyed that "all round, it is a no man's land of the ages, a stir-about of epochs and races, barbarisms and civilizations, virtues and crimes." He was a stickler for authenticity in his writing. "Nothing more shook his admiration for Herman Melville," recalled Stevenson's cousin, Graham Balfour, "than that writer's inability to approximate . . . the native names of the Marquesas and Tahiti." Among what Stevenson called the "farrago" of raw story material awaiting him in Honolulu, he heard about the tale of the castaways who came on the *Norma* two and a half months after the Stevensons arrived. "I am going down now," he wrote to his friend Sidney Colvin, "to get the story."

Stevenson recruited his stepson Lloyd Osbourne to search for more about *Wandering Minstrel*. Some sources—though not directly from Stevenson or the Walkers, and not verifiable—suggest Stevenson accepted an offer to read a diary Elizabeth kept on Sand Island, and that he judged it "one of the most thrilling stories ever written." The closest evidence that a Stevenson family member actually spoke to a member of the Walker family comes from a later newspaper report that Frederick met with Stevenson's son-in-law Joseph Strong, which was probably an error and meant to name Stevenson's stepson, Lloyd Osbourne. Stevenson's wife, Fanny, mentioned the *Minstrel* story "as told by the captain of the wrecked ship," which has been interpreted as meaning that Robert Louis and Fanny Stevenson met Frederick Walker, though she more likely referred to Frederick's press interviews that the Stevensons read along with the rest of Honolulu.

When the Stevensons toured other Pacific islands months later, they met John Cameron, who had continued working in the South Seas as a captain of blackbirding vessels. Upon hearing that John had been the first mate on *Wandering Minstrel*, Stevenson peppered him with questions. During the conversation, John drank heavily. He responded to questions with an air of "sly caution . . . as though he feared being trapped into some dangerous admission." He "vehemently" denied the fact that he had not reported the other castaways upon reaching safety with Hans Jorgensen and Mouwen, and insisted he had dutifully delivered Frederick's letter to authorities. Regarding John's claims, Fanny Stevenson correctly suspected "very little, if any, of it is true." In his usual manner of convoluted obfuscation, John explained that he stored comprehensive notes about the shipwreck locked in a safe on Jaluit, and handed the Stevensons a signed order authorizing them to be provided that manuscript—which of course did not exist. John's misdirections aside, Stevenson and Osbourne continued to fixate on the events and formulate the story for the novel that became *The Wrecker.*

EVERYONE INVOLVED IN the Midway wreck was changed by it. Freddy Walker spent significant time in the hospital, and after his physical recovery, the trauma lingered. From that point on, "every incident of that voyage [was] vivid in his memory." In particular, he thought about those who died after John Cameron and Hans Jorgensen failed to report their location. Because the treachery was more pointed coming from their own first mate than from a fugitive murderer, Freddy laid "the deaths directly at the door of Cameron."

In addition to the nine documented deaths related to the shipwreck of the *Wandering Minstrel,* which included those lost at sea in the stolen lifeboat, as well as Edward Dawson's death on board the *Norma,* an account in the *Pacific Commercial Advertiser* of Honolulu in 1900 indicated that a murder occurred during their time on Midway. Credence is added by the fact that Henry Walker, in the years after the family settled in Hawaii, worked for the *Advertiser.* George Hanker, who talked about the shipwreck to all who listened, spoke vaguely of this murder. A much later report from the U.S. Navy drawn from naval documents and records, published in Annapolis, Maryland, in *United States Naval Proceedings* in 1940, concurred, mentioning "several deaths from malnutrition, one from drowning, and at least one murder."

These references remain a mystery. There is no comprehensive list of crew members on board *Wandering Minstrel* as there would be for a naval vessel such as USS *Saginaw*, making it impossible to compare a roll call of those who boarded the *Minstrel* with those who left Sand Island alive. If John Cameron or Hans Jorgensen had committed this unidentified murder, the Walkers surely would have made it public when exposing their actions. Conversely, if one of the Walkers or the crew that remained under their protection had murdered someone, John would have made this known to justify his behavior. That suggests that a killing, if it happened at all, came to pass during the approximately five-month period after John and Hans departed from the atoll. Another clue may be hidden in the following language written about Frederick Walker's conduct by the naval court held by the British consulate in Honolulu: "No sufficient evidence exists to justify the Court in regarding the death of John Cooper as being from other than accidental causes."

No reference to a John Cooper has come to light anywhere else in extant material related to the shipwreck. The wording from the court's report does not suggest Cooper died from disease or malnutrition, nor does it fit with him being the unnamed crew member who drowned while attempting to fish. Rather, the language evokes a death that occurred in an ambiguous scenario, and could potentially be the same incident characterized elsewhere as murder. Nor does the court's wording rule out Cooper being one of the opium smugglers on *Norma*'s crew, rather than a sailor from the *Minstrel*.

To the extent that Robert Louis Stevenson and Lloyd Osbourne unearthed details otherwise lost to history, they may have threaded additional clues into *The Wrecker*. In that fictionalized version, the demand from the arriving captain that the castaways pay him to be rescued leads to violence breaking out between the two crews. "The suddenness of the attack and the catastrophe, the instant change from peace to war and from life to death, held all men spellbound." In the novel, deaths follow. If there was violence between the individuals of the *Wandering Minstrel* and the *Norma*, given the heightened emotions and the extortion by Charles Johnson, there may have been a tacit agreement later between the groups to stay quiet about any resulting death, similar to their agreement to publicly state that the Walkers "chartered" the *Norma*.

The Walkers recuperated after their rescue while residing in Puuloa (Pearl Harbor). "Our experience there," Frederick wrote, "was that we could sleep 20 hours, eat three, and amuse ourselves in the water the other one hour of the 24." Jessie could frolic with them, scampering in and out of the water. Five months later, they moved to Honolulu. They ended up forming a deep attachment to Hawaii and a lasting gratitude for the efforts of new friends to help them after they had arrived, for all intents and purposes, insolvent. The possibility arose of returning to Midway with another ship to salvage what they could from the *Minstrel* wreckage, but Frederick was infuriated to hear that Charles Johnson sailed the *Norma* to Midway two months after the rescue to do just that.

Meanwhile, the Walker boys still grappled with what John and Hans had done. In England, the Jack the Ripper murders—all of which occurred during the period when the Walkers were on Sand Island—represented to the world the horror of violence coming from a person who blended into the crowd. But what the Walkers had faced was confronting darkness in someone in a position of trust, more akin to Stevenson's Dr. Jekyll and Mr. Hyde character. Henry Walker burned with resentment that the family heirloom compass Frederick had given John Cameron was never returned, a small but material symbol of his treachery.

Rather than wait in Honolulu for some word about their

adversaries, the youngest Walker, Charlie, decided to take matters into his own hands and sail to the Marshall Islands to investigate. Speaking to authorities there, he was able to reconstruct the arrival of John, Hans, and Mouwen at Jaluit, and to discover how John and Hans concealed where they had come from. Charlie also uncovered the story of Hans being sentenced to death for his crimes against an islander before being rescued by the *Montserrat*, which carried Hans to Central America. (The *Montserrat* itself vanished without a trace in a hurricane two years later, killing everyone on board.) Charlie learned that while living in Guatemala, Hans was killed in a brawl.

At one point, John Cameron faced his own reckoning. He was detained at the Jamestown Harbor in the Caroline Islands by a Spanish warship called *San Juan de Ulloa*, also the name of a notorious island prison. John was placed on trial for illegally selling arms. However, an ally of John's used extrajudicial means—bribery or favors—to secure his release. Later, John was spotted captaining his blackbirding vessel the *Ebon* flying a black pirate flag. John had taken possession of the *Ebon* from Crawford & Company in lieu of money they owed him. When the *Ebon* sailed under a black flag, she crossed paths with a bark helmed by Captain J. F. Luttrell, the fellow blackbirder from whom John reportedly had absconded with $1,600 of goods years earlier. Luttrell said "he would have given one of his eyes to have shot" John. The following year, after eluding the vengeful Luttrell, John sailed the *Ebon* to Kobe, Japan, carrying a crew of very ill Pacific Islanders. A member of the crew described as a boy died shortly after arrival. Three other crew members died from smallpox after days of suffering. Meanwhile, John fled Kobe on another vessel, having taken a commission to bring Russian soldiers to Vladivostok.

At strategic points, John continued to change his story about what happened at Midway, turning himself into a hero (comparing himself to Odysseus) and the Walkers into villains, even absurdly alleging that Frederick intentionally wrecked the ship, and that John had sailed on the *Minstrel* on orders of the British government to spy on the Walkers. He described Hans Jorgensen as "a good-natured, obliging and harmless individual." John had an almost pathological need and ability to persuade skeptics to his side, using what the Walkers called his "tissue of falsehoods," an intricate combination of accurate details and brazen fabrications. He managed to win over a maritime writer to help him spread his disinformation, after that same writer had asserted that John and Hans Jorgensen should be hanged for what they had done to their *Wandering Minstrel* shipmates. John Cameron never faced punishment for what happened in the shipwreck, nor for his bloodthirsty acts as a blackbirder.

Mouwen, who had fled from Hans and John, ended up joining an American trading schooner called *H. L. Tiernan*, which capsized near the Gilbert Islands on November 4, 1889, resulting in eleven people drowning. Mouwen's fate remains unclear, but evidence suggests he was rescued.

The array of insurance agents, police officials, and other investigators who had chased John and Hans moved on to other cases. Within a year of the rescue of the Walkers, the once prosperous San Francisco branch of the South British Fire and Marine Insurance Company, one of *Minstrel*'s insurers and pursuers of John Cameron, faced hard times. Alexander Murray, the firm's local manager, discovered that his cashier Frank Agnew had embezzled or, in a charitable interpretation, misplaced thousands of dollars, which was investigated by the Harry Morse Detective Agency. On top of this, payouts for shipwrecks including *Wandering Minstrel* mounted. The corporate headquarters in New Zealand shuttered the branch, and Alexander and his secretary, Charles Stringer, were reassigned abroad.

Long after the unsuccessful search for John Cameron, detective Harry Morse continued taking high-profile cases. One of the most prominent, the poisoning death of Jane Stanford, former California First Lady and part of the founding family of Stanford University, remains unsolved. Morse parlayed his notoriety into endorsements, including one for a rheumatism medicine prepared at an Oakland pharmacy, which the advertisement noted could be trusted because "who does not know the ex-Sheriff of Alameda County whose career has been as brilliant as it was unblemished?" The remedy likely contained opium, which would have been imported through legal means that Morse helped protect by hindering smugglers like John Cameron and corrupt government accomplices.

USS *Alliance*'s Henry Clay Taylor, who had declined taking action against John Cameron after inspecting his ship at Kusaie, three years later became the president of the Naval War College, before being promoted to rear admiral. Edward Davis of HMS *Royalist*, the more zealous investigator who attempted to bring Hans Jorgensen to justice for crimes on Pacific islands, was soon after given a post in England, where he married and settled in Sussex, and had four sons. He commanded various fleets and battleships before retiring.

The steamship captain William Seabury, who came close to anchoring at Midway, later ran across Freddy Walker in Hawaii. Seabury described to Freddy how he had feared wrecking the *City of Peking* on the reef. The concern proved justified, as another Pacific Mail steamer wrecked there years later, around the same time that Seabury died in an accident in Fall River, Massachusetts.

Captain Henry Stap, who had nearly sought shelter from storms at Midway in the merchant ship *Halewood*, ended up having a connection to the most famous shipwreck in history. His daughter Sarah was a ship stewardess who survived the sinking of the *Titanic*. The *Halewood* was sold to a Norwegian firm and rechristened *Songvig*. In 1920, the ship hit a mine, a growing peril at sea, and twenty-six people on board died.

Another ship that could have rescued the Walkers, USS *Juniata*, broke down the year of the castaways' rescue. Commander George T. Davis retired in the coming years, but insisted on coming out of retirement when the Spanish-American War began. Meanwhile, fellow Pacific squadron naval sloop *Vandalia*, whose crew trounced the *Juniata* at baseball while in Honolulu, was sent to secure American interests by countering the German presence in Samoa. The same day that the castaways on Midway began boarding the *Norma*, the *Vandalia* encountered a powerful cyclone and crashed against a reef, leading to forty-three casualties.

The vessel that closed in on Midway during its plot to stage a shipwreck for insurance money, the clipper *Brussels*, was sold to a new owner in London who overlooked its "checkered history" and funded repairs after the fire at San Francisco harbor. Attention on the case faded after the arson convictions of first mate Clarence Caskey and steward Charles Bachman, but questions remained about the potential complicity of Captain Robert Crosby. Crosby had not had his final encounter with law enforcement. While in charge of the *Baron Belhaven*, Crosby attempted to leave Bangor, Maine, without proper clearance for his cargo. The U.S. revenue cutter *Woodbury* lay in wait at the mouth of Penobscot Bay, with cannons loaded with twice the usual ammunition, prepared to damage or even sink the *Baron* in order to stop Crosby. Heavy fog stalled a confrontation, leading the *Baron* to run aground, after which the armed crew arrested Crosby. In a jumbled explanation, Crosby claimed there had been a misunderstanding because jewelry and other valuables had been stolen from him, which he had also alleged during the *Brussels* case.

The only captain to successfully reach the castaways, Charles Johnson, ultimately collected more than a thousand dollars in payments for rescuing the Walkers and their crew, a lower sum than he wanted, particularly after the *Norma*'s owner claimed a share of the money. Charles continued to look for opportunities to smuggle opium, taking command of a notorious ship in that trade called *Halcyon*. Unable to resist easy money, Charles stole a five-dollar gold piece from the pocket of a drinking companion in Port Townsend, Washington Territory. While in jail, Charles bragged to an undercover reporter about his prolific opium smuggling. His rescue of the castaways of the *Minstrel* remained the most lauded achievement of his career, with his quasi extortion of the Walkers mostly forgotten over time. Returning to the helm of *Norma* three years after rescuing the Walkers, Charles blew himself up while using explosives for an unknown purpose.

Charles's former first mate, John Forbes, who had stolen Charles's earlier ship, the *Nereid*, disappeared without any indication of whether he located the treasure Charles had been searching for. A Spanish explorer publicly claimed to have found it. Still, spots across the Mariana Islands were later dug up by an expedition looking for the treasure, which remains unaccounted for to this day. Charles may have missed a divergent legend that the original pirates who stole the ship and fortune from Peru buried part of the treasure in Oahu, Hawaii, where, a later historian suggested, it presumably remained "waiting for a treasure hunter to come and find it."

Frederick Walker still dreamed about finding a different treasure, the one he had heard about from the old sailor who talked about gold and silver buried by pirates on Palmyra Island after a shipwreck on the reef. That wreck scenario no doubt now reminded Frederick of the end of his own voyage to Midway. Frederick apparently discovered a morbid curio in the form of a map of the treasure's location that was drawn on human skin, but he never found the treasure.

A legend spread that within a couple years of the *Minstrel* rescue, most of the former castaways had died, as though beset by a curse, but Frederick maintained a sense of humor when it came to the topic of their shipwreck, calling it their "unpleasant picnic of fourteen months' duration." Still, he had come to believe for the first time in unlucky ships, though was undaunted about the prospect of returning to the site of their disaster. His celebrated conduct during the crises of the *Minstrel* led to new professional opportunities. The government commissioned Frederick to take the schooner *Kaalokai* to survey area islands, an itinerary that included Midway. Freddy, now twenty-one, and Charlie, seventeen, decided to join as their father's mates. A group of naturalists traveled on board to carry out a study of birds. Of all the islands visited, one of the naturalists found Midway "the most forbidding and weird." They also had to suppress their concerns about sailing to the place where the Walkers had wrecked, as if the island wanted the family back. There was some reason for this superstition. When the *Kaalokai* stopped at the French Frigate Shoals, the Walkers found a green plank that they identified as part of the *Minstrel*—having floated from Midway more than five hundred nautical miles to their exact location two years later.

Even before they stepped back ashore at Sand Island, the Walkers could spot the hut in which they had sheltered. "There was the old house," Frederick commented, "still standing!" He wrote out a log of their observations as they sailed closer. "Stood on to westward past the barrier reef, then hauled to the southward till the channel or passage was open; then hauled in and worked up to the anchorage." They anchored the *Kaalokai* in Welles Harbor where the *Wandering Minstrel* had wrecked.

"We felt a sickening feeling in our throats as we landed," Frederick admitted. The three Walkers explored their so-called village, the condition of which had predictably deteriorated, with the mutton birds swarming the shelters the castaways had counted on for survival. The door to their main hut had been blown off. When a passing schooner anchored, Frederick showed the visiting captain inside the hut. The captain rushed right out, yelling, "I could not stay there! It feels like a living tomb!"

One of the naturalists also decided to enter the hut. He tried to imagine the Walkers' experiences inside. "There is something melancholy about this desolate place; the sigh of the wind round the house, the wail of the petrel, at any time melancholy, seems even more so, and looking out of the door over the extent of white sand broken only by a few green mounds and the little group of crosses in the cemetery, a feeling of depression comes over me, the smell of dry goonie meat and the thought that a woman had to exist there all that time and in such company all tend to the weirdness of the place." The outsider assumed Elizabeth was helpless among the castaways rather than a leader.

Freddy, meanwhile, cultivated an interest in photography and memorialized their return with a series of portraits of Midway, capturing views where "from horizon to horizon nothing can be seen but sea and sky." One of the photos revealed a landscape of hundreds of birds, which Freddy wryly titled *A Few Tern*. He took a shot of the shelters that he called *The Deserted Village*.

The naturalists complained to Frederick about remaining too long on Midway. But for the Walkers, there was a strange appeal to reliving their experiences, as though regaining control over them. Taking a smaller boat out in the bay, they dropped a glass-bottomed box over the side, through which they could examine the seafloor. They spotted chains and one of the *Minstrel*'s lost anchors. Charlie sighted something else thirty feet below—the old cannon that had enchanted the Walker boys in a simpler time, when they first boarded the *Minstrel*.

They had traveled with divers on board, but the divers were afraid of the sharks. "I decided to chance it myself," Charlie, the born leader, recalled later. Charlie stripped, putting on a belt with a sailor's knife in it. "Going down with a length of heavy rope, I reached the cannon. I had just tied a noose around its muzzle when an ominous shadow passed by on the sunlit floor of the sea." The dark shape revealed itself as a shark, so integral to their original voyage's purpose and ultimately to their survival when they had to feed starving family members. "Those in the boat above saw the shark at the same time. I started up, pulling with mighty, frantic strokes toward the surface." The twenty-foot shark circled back toward Charlie. "I reached [the] surface just in time and waiting hands pulled me into the boat." They dragged the cannon into the boat. It became a fixture at the Walker homestead in Honolulu, though Charlie shuddered whenever he saw it, thinking how he nearly died retrieving it.

The Walkers soon prospered enough to acquire another important relic of their experiences: the *Norma*. The vessel had been offered for sale after the explosion that killed Charles Johnson.

Elizabeth never showed any desire to return to Midway. Shortly after their rescue, she had learned about her ill sister-in-law, Kiyo Eyton, consulting a soothsayer in Japan, who assured Kiyo that the Walkers lived at a time when everyone else insisted they had perished. Kiyo died the year after their rescue, at age forty. J. L. Eyton remained in Yokohama and remarried. Elizabeth welcomed the idea that the soothsayer had some kind of authentic spiritual vision, and did not dispute a quasi-religious interpretation that her family's survival ultimately showed "victory through faith."

ROBERT LOUIS STEVENSON'S determination to publish a version of the events on Midway contributed to his family's decision to settle in Samoa, which had a busy enough port to allow Stevenson to transport manuscript pages to a publisher. Living in Samoa became a turning point for the Stevensons. While opposing the control of Germany, the novelist established a kind of colonial fiefdom, turning himself into a figure resembling the character of Kurtz in his literary successor Joseph Conrad's novel, *Heart of Darkness*—in fact, anecdotes about Stevenson's transformation were likely relayed directly to a thirty-five-year-old Conrad, who was sailing through the Pacific at the time as a first mate.

Upon first hearing about the *Wandering Minstrel*, Stevenson perfectly summarized the unique narrative tension: "a shipwrecked family . . . on an island with a murderer: there is a specimen." However, as he and his stepson shaped and wrote *The Wrecker*, the family element vanished in favor of featuring convoluted dynamics between multiple crews that wreck or anchor at Midway, resulting in a sprawling, sometimes confusing plot that included characters resembling Hans and Charles Johnson. The book was so far removed from what made the real events most compelling that nearly a decade after its publication a newspaper column in Honolulu referenced the *Minstrel* as though *The Wrecker* did not exist, noting "its details would furnish material for a novel that would be worthy of the efforts of a Stevenson or a Defoe." Still, *The Wrecker* became one of the more successful titles in Stevenson's oeuvre, helping to finance the novelist's eventful final years in Samoa.

Secretary of State Thomas Bayard was lampooned when he left office for his lack of aggressiveness toward other nations, with one commentator sniping that "he ought to be pensioned by the various foreign governments which he has served at every opportunity during his political career." When it came to the Pacific, he had remained uncomfortable asserting America's rights on Midway or beyond. Other officials and power brokers from the United States held no such qualms, continuing to consolidate control over Hawaiian political and military life. When disgruntled rebel Robert Wilcox returned to Hawaii from his exile, he led yet another attempt against the "American plunderers" in his latest ill-advised plan to install the princess to power. Three months after the arrival of *Norma* with the rescued castaways, Robert occupied the 'Iolani Palace in Honolulu with cannons and rebel soldiers. He was arrayed in the Italian military uniform he had received during his studies abroad. He embraced a quote he attributed to Napoleon: "Bravery is the motto, and this is my seal as a professional soldier." Hay Wodehouse, son of the British commissioner who had overseen the 1887 departure of the *Wandering Minstrel*, was called upon for his baseball pitching skills to toss bombs on the bungalow from which Robert was commanding his charge. "His enthusiasm was great," the Hawaiian princess recalled of Robert's plots, "but was not supported by good judgment or proper discretion." Robert's legacy remains complex, with a statue standing today said to represent his fight for "truth, independence, nationhood, and sovereignty in a day when American imperialism was conquering the Pacific."

American political, business, and commercial leaders in Hawaii commanded their own military maneuvers that culminated in the overthrow of the monarchy, the declaration of a republic, and ultimately the annexation of Hawaii by the United States. The statement of the naval officer who first claimed Midway for the

United States could be recalled with new significance: "I sincerely hope that this instance will by no means be the last of our insular annexations."

In hindsight, Postmaster Frederick Wundenberg sending a gift basket onto the *Norma* may have been more than a greeting for the rescued Walkers and their shipmates, as it also served to celebrate that an American captain, Charles Johnson, had successfully anchored at Midway. Wundenberg would prove adamantly in favor of American influence and ultimately absorption of Hawaii. The *Norma*'s American captain had inadvertently carried out some of the symbolic activity that the *Juniata* had failed to do a year earlier. By demanding payment from the Walkers, the greedy *Norma* captain had even given an example of an American conducting business at the atoll, which could bolster the United States' argument for continuing to use Midway for commerce, an element supporting a possession claim. Midway had been one of the first offshore islands annexed by the United States, paving the way for more.

Frederick complimented Hawaii's "genuine hospitality to worthy strangers (and very often to very unworthy ones)." Frederick and the Walker sons were viewed by peers as loyal to their adopted home of Hawaii. This partially backfired for them as rebels disrupted American control of Hawaii. At one point, the Walkers were accused of using the *Norma* to secretly transport weapons within a purported shipment of coconuts, to arm revolutionaries in Honolulu who wished to restore the monarchy. Frederick denied the allegations, and after some misleading reports that he was deported, he returned to Hawaii. A real movement for restoration arose, including yet another unsuccessful attempt by Robert Wilcox. James Wodehouse vocally sided with the deposed royals and Hawaiian self-rule—or at least against American rule. But battle lines were not obvious in Hawaii. Another diplomat, Henry F. Glade, the

German consul whose reprimand of Hans Jorgensen had led him to join the *General Siegel*, was part of a committee instrumental in the overthrow of the royal family.

Edvart Olsen, one of the survivors of the *General Siegel*, worked for the Oahu Railway & Land Company, which was founded shortly after Edvart's passage to Hawaii from Midway. The railway allowed the sugar plantations to prosper by providing access and transportation. Edvart ended up making further impact on Hawaiian history by serving in the militia that contributed to toppling the monarchy. The Norwegian Edvart followed the pattern of former castaways recognized (and exemplified) by Frederick Walker for embracing life in Hawaii, regardless of their politics. "Many, like myself," Frederick remarked, "are sea waifs, rescued from shipwreck, brought here and declined to move on, but commenced life anew, and are now well satisfied with their decision." Frederick noted that descendants of the castaways from the double shipwreck of the *Pearl* and *Hermes* in 1822 had become some of Hawaii's most successful and influential residents.

The Walkers remained involved with Midway. Frederick corrected charts and maps used for navigation for years afterward. Upon their return to Sand Island, Frederick, Freddy, and Charlie also dug a better well for future visitors, and within a few years windmills were added to raise water from the well. Freddy had released birds called Laysan rails, which could not fly, onto Green Island during the *Kaalokai* cruise, and their population blossomed over the ensuing years. The U.S. Navy returned to Midway, too, sending steam sloop USS *Iroquois* to prepare the atoll to serve as a key transmission point for telegraph cables that would allow a direct communication pathway between the United States and Asia. Conducting thousands of soundings around the atoll, the *Iroquois* came across a 1,200-pound anchor lost by the *Wandering Minstrel*, which they returned to Honolulu. Project leaders also consulted with Frederick Walker about how to best manage the features of the two islands of the atoll. With the resulting telegraph line, President Theodore Roosevelt sent the first message to be transmitted around the globe. Amid expanded military and commercial operations, at one point Sand Island boasted a full-size golf course.

Not long after the first "around the world" telegram was cabled, SOS became an emergency Morse code signal for ships in distress. The two Ss and the O signify numbers of dots and dashes, which could be interpreted by the receivers of the signal, and were not intended as abbreviations. However, people associated SOS with "save our souls" or "save our ship." The phrase "save our souls" appears in the much earlier quote about shipwreck from an 1869 Victor Hugo novel chosen as the epigraph of this book. The phrase succinctly captured the profound themes that emerged in the narratives of the stranded, both in fiction and in real life, when there are stakes not just of survival but of humanity—the key difference, as Hugo put it, between what it meant to "save our lives" and "save our souls."

Elizabeth and Frederick died within a year of each other at seventy-one and seventy-eight years old, respectively. Though involved in community and business, the Walkers and especially their sons tended to wall themselves off from the world in the years after the shipwreck. Of the three, only Freddy married, and after an acrimonious divorce, his ex-wife and his only child, Violet, moved to California. The three Walker sons remained a tight-knit unit. Charlie designed and built a three-story structure, with the words "Elizabeth Building" inscribed upon it. The building was an unintended echo of *The Swiss Family Robinson*, in which the fictional family constructed a boat they christened *Elizabeth*. Freddy's photography was among the first to publicize "Hawaii's scenic beauty," and in addition to his photography career he also invented equipment to raise sunken ships. Charlie, who became a master shipbuilder, preserved relics of their family's experiences, including his father's human skin pirate treasure map and the cobweb-covered *Wandering Minstrel* anchor salvaged by *Iroquois*. The three brothers lived together in the Elizabeth Building and operated a boat and carriage repair shop out of the ground floor, as well as a boatyard a few hundred feet away, while Henry was aided by his brothers in a business importing and milling rice—an echo of the life-saving cask of rice that washed ashore during their time as castaways—in between his competitions in bicycle and yacht races. Representing their own kind of island amid bustling Honolulu, the Elizabeth Building stands today. Newspaper retrospectives about the *Wandering Minstrel* sometimes mistakenly mentioned four Walker sons, probably a confusion with the fictional Swiss Family Robinson.

Shipwrecks make an impact beyond the individuals involved, changing and often becoming part of the topography and ecology of atolls, reefs, and islands, creating new ecosystems. Some islands' very identities depended upon wrecks, with the Pearl and Hermes Atoll deriving its name from the two early shipwrecks there. The wrecks of *Dunnottar Castle* and the *Saginaw* at Ocean Island, now called Kure Atoll, have been discovered and continue to be studied. Some wreck sites, including *Dunnottar* in 2006, were discovered by chance and are only accessible with the best equipment under certain conditions. Midway Atoll presents special challenges because of the depths around the reefs, and neither the remains of the *Wandering Minstrel* nor the *General Siegel* have been found.

Within a decade of the *Minstrel* wreck, sailing schooners in professional and commercial pursuits were being overshadowed by steam-powered vessels, considered by many to be safer and more reliable. In the single month in 1889 when the Walkers were rescued, the insurance industry counted eighty-six sailing ships lost (including wrecks, collisions, and fires), in contrast to only fifteen steamships in the same period. (Harder to come by would have been accurate numbers of how many vessels of each category were at sea.) These steamships were larger and carried out multiple purposes—transporting cargo, mail, passengers—in a more organized manner. Sailing vessels still had a place because of the cost of coal and the complex logistics needed for steamships, and engine power was not without detractors. Oscar Wilde found his ocean travel on the steamer *Arizona* (the same ship that passed along John Cameron's fabricated story of his heroism) fell short of what he imagined and lacked any romantic quality. "It was very tame. I expected to have it roar about and be beautiful in its storms. I was disappointed in it."

As for sharking, newer vessels in the Pacific in the late nineteenth century overfished to such a degree that the industry became practically defunct for a time by the early 1900s because of the scarcity of sharks. The financial uncertainty in sharking continued to render opium smuggling a bigger temptation to those vessels' sailors—as exemplified by the *General Siegel* and the *Norma*—filled with its own dangers and detriments. During the time of the *Wandering Minstrel*'s wreck, the Walkers and the other sharking vessels had to kill more sharks to ensure profitability and had to become bolder, including fishing during life-threatening weather. In storms, some big sharks came closer to the surface to hunt, making them more likely to be caught. The Walkers' year-plus on Sand Island can be seen as a parable for the consequences of environmental catastrophe, a tableau of humankind left in a barren environment with dwindling resources. The search for accountability and ethics in humanity's interaction with sharks continues to this day. Late in his life, Frederick almost died from illness and attributed a dramatic recovery to sharks, specifically to ingesting their liver oil.

The concept of a deserted island captured the notion of land as a blank slate, a place to remake a person's identity, relationships, and future prospects. This also held a mirror to the way colonial powers used money, power, and militaries to try to turn the Gilbert, Marshall, and Hawaiian islands, among many others, into blank slates—despite the fact that they were very much inhabited.

Montgomery Sicard (later promoted to rear admiral), who as captain of the *Saginaw* oversaw the attempt to widen passage through Midway's reef, was honored posthumously when destroyer USS *Sicard* was launched by the U.S. Navy in 1920 to engage in patrol and transport duties. The *Sicard* was at Pearl Harbor when the naval base was attacked in 1941 and aided the fleet. The destroyer for a time anchored at Midway's Green Island, by then known as Eastern Island, to establish a communications center prior to the Battle of Midway, which became a turning point in World War II that brought one of the most remote spots on the map, which once befuddled the *Sicard*'s namesake, to the world's attention. The Midway Atoll that was once conceived as a coal refueling station for steamships became a crucial refueling station for submarines.

The number of birds on Midway has continued to astonish observers, and by some accounts, audio that was recorded on Midway served as sound effects for Alfred Hitchcock's 1963 film, *The Birds*. The reign of the seabirds on Midway has been consistent but not without disruption, especially from rodents. Crew members of both *General Siegel* and *Wandering Minstrel* reported mice and rats on Midway, very likely transported to the middle of the Pacific on their own schooners. U.S. Navy vessels in the 1940s brought populations of black rats, which attacked and killed all the flightless Laysan rails on Midway (these birds were descended from those released by the Walkers upon their return to the atoll on the *Kaalokai*). Having disappeared from Laysan Island decades earlier, the loss of Laysan rails on Midway meant extinction. An estimated fifteen thousand rats were killed early in 1945 in an effort to protect bird species, and were believed to have been eradicated by 1996 after a series of pest control initiatives, but invasive rodents have continued to plague the atoll's birds, including an unusual acquired behavior in which house mice began feeding on live adult albatrosses.

Midway is littered with plastic and other human-made debris that washes from the ocean onto the shore, from thousands of miles away. Albatrosses also unwittingly feed their young pieces of plastic that kill them. Decaying bodies of birds glimmer with bottle caps, and the sand itself slowly changes composition as particles of plastic become part of the environment.

Honolulu-born president Barack Obama visited Midway in 2016 after declaring the world's largest protected marine reserve around the atoll. With climate change leading to scientists warning that sea levels can rise fifty feet by the year 2300, the atoll, with Sand Island's highest elevation at fifty-seven feet above sea level, could eventually disappear, likely long before the plastic decays. The president toured the island, where a population of between forty and fifty people oversee conservation efforts. The presidential delegation would have crossed through the same sites where the Walkers and their fellow castaways fought to survive.

Those who perceive Midway as having some kind of mysterious pull could add to the list the tale of the Taiwanese fishing vessel *Yong Yu Sing No. 18*, which was discovered adrift near Midway on

January 2, 2021, with no sign of its fifteen-person crew. The so-called "ghost ship" showed no traces of a struggle, giving rise to a theory that all crew members were swept into the sea.

In a preface to *The Wrecker*, Fanny Stevenson wrote that "my husband tried in vain to solve the mystery of the *Wandering Minstrel*." For Stevenson, that mystery was the unorthodox arrangement between the Walker family and Charles Johnson. But other mysteries and contradictions persist. There is the question of the possible murder and the death (possibly one and the same) of a man named John Cooper. Contradictions also remain about some details, such as whether the first contact between the *Minstrel* and Hans Jorgensen was when he stepped outside the hut, or whether Hans was first invited to board the schooner when it anchored in the harbor before wrecking. There are also outlying suggestions that Frederick chose to chart the *Minstrel*'s course to Midway for political reasons. One account seven years after the shipwreck claimed Frederick had sailed to Midway intending to claim it in the name of Great Britain. In contrast, a later report on the wreck by the U.S. Navy referred to the *Minstrel* as having been commissioned by the government of Hawaii to survey Midway, which would fit with the inklings of Honolulu officials wishing to claim possession of Midway at the time, but would require a perplexing conspiracy to make sense of the fact that no rescue vessel was sent when the schooner failed to return.

Over the years, Elizabeth Walker opened up about her time as a castaway when asked by visitors. She mulled different mysteries of the shipwreck than Stevenson had. She considered how striking it was that Freddy's health had turned on something as small and unlikely as a bottle of lime juice found in the ocean. She also pondered the fact that the physically strongest crew members ended up being the most violent and greedy among them. The most enduring conundrum might remain the one that the castaways themselves confronted at the time: whether people torn from civilization and forced to rebuild their world from virtually nothing survive by turning against or toward each other.

With the vines above casting dappled shadows onto the table of the Walkers' *lanai*, or veranda, in Honolulu, Elizabeth would pour tea from the battered silver tea set that had washed up onto Sand Island after the wreck. As she sipped from her cup, she reflected how "storm and stress seem very far away."

AUTHOR'S NOTE AND ACKNOWLEDGMENTS

AS THE FIRST BOOK-LENGTH ACCOUNT of the *Wandering Minstrel* shipwreck, I was fortunate to be able to consult a wide range of sources, a process that included help from the National Archives of the United Kingdom, which holds documents assessing the arrangement between the Walkers and Charles Johnson, and the Special Collections of the University of Adelaide Library, which assisted with material related to Captain Edward Davis and HMS *Royalist*. Dr. Steven Park generously lent his expertise in maritime history in reading my manuscript, Nick Ripatrazone shared his Biblical knowledge, and George Glazer advised on historical globes and maps.

When a quote conveys written text—or carved text, in the case of the messages on the walls of the Sand Island hut—the original capitalization, spelling, and punctuation are preserved, while I occasionally normalize those aspects in quoting spoken words.

Throughout work on this book, I have once again relied on my agent, Suzanne Gluck, and my editor, Sara Nelson, for their wisdom and insights, alongside the WME team, including Andrea Blatt, Peter Dunn, Lane Kizziah, and Hilary Zaitz Michael, and the HarperCollins team, including Jonathan Burnham, Edie Astley, Nikki Baldauf, Michael Siebert, Michele Cameron, and Jane Cavolina. My extended circle always ready with advice and encouragement, in writing and otherwise, includes Susan and Warren Pearl, Marsha Selley, Kevin Birmingham, Benjamin Cavell,

206 AUTHOR'S NOTE AND ACKNOWLEDGMENTS

Gabriella Gage, Joe Gangemi, Jon Housman, Greg Nichols, and Scott Weinger. My wife, Tobey, read and improved this book in real time, and my daughter, Lulu, gets bonus credit for suggesting the perfect title; Cooper, Graham, and Lulu, you always make us proud, and we know, if stranded together on a deserted island, you would find us a way home.

NOTES

Prologue

1   With time running out: "Report of Wreck," National Archives of the United Kingdom, BT 15/30/F4673; Wreck Report, "Wandering Minstrel," Port Cities Southampton, No. 3792; Elizabeth Walker, "Wrecked on Midway Island in 1888," *Paradise of the Pacific*, November 1936, 27; John Cameron, transcribed by Andrew Farrell, *John Cameron's Odyssey* (New York: Macmillan Company, 1928), 276.

1   The chains of the anchor: Frederick D. Walker, *Log of the Kaalokai* (Honolulu: Hawaiian Gazette Co., 1909), 55; Epes Sargent, *Songs of the Sea* (Boston: James Monroe and Company, 1847), 50.

2   Three lifeboats were lowered: "Sons of Captain Walker Refute Wandering Minstrel Story," *Honolulu Advertiser*, January 30, 1922.

3   The Walkers now had to rappel: "Survivors Now Here," *Honolulu Advertiser*, August 15, 1900.

Chapter 1: The Voyage Begins

5   Six months earlier: Church of St. Peter (Liverpool, Lancashire County), Register: Baptisms 1851–1852, 141: 1124; Herman Melville, *Redburn* (New York: Harper, 1849), 209.

5   Elizabeth had married: "Married," *Liverpool Mail*, May 11, 1867; "A Honolulu Man Knows Where Much Silver Has Been Buried," *Honolulu Advertiser*, July 6, 1903; Frederick D. Walker, *Log of the Kaalokai* (Honolulu: Hawaiian Gazette Co., 1909), 19; "The Shark Fishery Company," *Overland China Mail*, February 26, 1890.

6   Hong Kong, a British colony: William Gifford Palgrave, *Ulysses: Or, Scenes and Studies in Many Lands* (London: Macmillan, 1887), 201.

6   Elizabeth was not the only woman: Abby Jane Morrell, *Narrative of a Voyage to the Ethiopic and South Atlantic Ocean, Indian Ocean, Chinese Sea, North and South Pacific Ocean, in the Years 1829, 1830, 1831* (New York: Harper, 1833), 17–18; Joan Druett, *Hen Frigates: Wives of Merchant Captains Under Sail* (New York: Simon & Schuster, 1998) (excellent study of women on sailing ships).

7   Elizabeth's role on ships: "Lady Whalers," *Whalemen's Shipping List and Merchants' Transcript*, February 1, 1853; "Forty-two Wives of Whaling Captain in the Pacific," *The Friend* (Honolulu), November 8, 1858; *Vancouver Daily World*, December 22, 1890.

7   By the time the Walkers: "The Shark Fishery Company," *Overland China Mail*, February 26, 1890; "An Unlucky Peterhead Vessel," *Peterhead Sentinel and Buchan Journal*, May 7, 1889; "Hurricane in Japan," *Peterhead Sentinel and Buchan Journal*, September 24, 1889; Walker, *Log*, 55.

8   The Walker boys knew: Correspondence from Laura Jernegan to her grandmother, March 6, 1870, Subgroup 19, Beetle Papers, Nicholson Whaling Collection, Providence Public Library, Providence, RI; Mary Chipman Lawrence, *The Captain's Best Mate: The Journal of Mary Chipman Lawrence on the Whaler Addison, 1856–1860*, ed. Stanton Garner (Providence, RI: Brown University Press, 1966), 3, 11, 43, 61.

9   Before they embarked: "Old Cannon, Taken from Pirate Junk, Is Recovered After Wreck at Midway," *Honolulu Advertiser*, April 22, 1923; Alexander MacDonald, "Down the Gang Plank," *Honolulu Star Bulletin*, March 6, 1934.

9   Frederick Walker was one: Maria de Oca Echarte, "Reef Sharks: A Historical Ecology and Media Project," Capstone Project, Scripps Institution of Oceanography, 2009; Robert Louis Stevenson and Lloyd Osbourne, *The Wrecker* (London: Cassell, 1892), 189.

10   The *Wandering Minstrel*'s name: Arthur William À Beckett, *The À Becketts of "Punch": Memories of Father and Sons* (Westminster: Archibald Constable and Co., 1903), 50; "A Wandering Minstrel," *Daily Bulletin* (Honolulu),

November 24, 1887; Fletcher Bassett, *Sea Phantoms: Or, Legends and Superstitions of the Sea and of Sailors* (Chicago: Morrill, Higgins & Co.: 1892), 443–46.

10   After the typhoon: Walker, *Log*, 55; "A Wandering Minstrel," *Daily Bulletin* (Honolulu), November 24, 1887.

11   At that time: James Mak, "Creating 'Paradise of the Pacific': How Tourism Began in Hawaii," *UHERO Working Paper No. 2015-1*, Economic Research Organization at the University of Hawai'i, February 2015, 13; Richard A. Hawkins, "The Impact of Sugar Cane Cultivation on the Economy and Society of Hawaii," *Illes i Imperis* 9 (December 2006): 68; Ralph Kuykendall, *The Hawaiian Kingdom, Volume 3, 1874–1893: The Kalakaua Dynasty* (Honolulu: University of Hawaii Press, 1967), 395.

11   But weeks later: Liliuokalani, *Hawaii's Story by Hawaii's Queen* (Boston: C. J. Peters & Son, 1898), 181–82; *Hawaiian Gazette*, September 20, 1887.

11   Frederick did not want: Lawrence, *Captain's Best Mate*, 169.

12   The first and second mates: John Cameron, transcribed by Andrew Farrell, *John Cameron's Odyssey* (New York: Macmillan Company, 1928), 241.

12   Wodehouse sent a local police deputy: "Three Runaways," *Evening Bulletin* (Honolulu), November 7, 1887; "A Find," *Hawaiian Gazette*, October 25, 1887.

12   Bracing for a physical fight: "Mutiny," *Daily Bulletin* (Honolulu), December 9, 1887; "Local and Central News," *Daily Bulletin* (Honolulu), December 10, 1887; Cameron, *Odyssey*, 241; "The Mystery Solved," *Daily Pacific Commercial Advertiser* (Honolulu), June 13, 1889.

13   With the crew was filled out: "The Wandering Minstrel," *Daily Bulletin* (Honolulu), April 8, 1889; Walker, *Log*, 15–16.

13   In a week: A. Binion Amerson, Jr., "Natural History of French Frigate Shoals, Northwestern Hawaiian Islands," *Atoll Research Bulletin* 150 (December 20, 1971): 40; Lawrence, *Captain's Best Mate*, 5.

13   The new second mate: "Death of George Hanker," *Hilo Daily Tribune*, December 13, 1901; "Wedded to Bad Luck," *Hawaiian Star*, December 17, 1901; George C. Munro, "Nature Studies," *Honolulu Star Bulletin*, May 2, 1931.

Chapter 2: The First Mate's Story

15   Writer Fanny Stevenson described: Fanny Stevenson, *The Cruise of the "Janet Nichol" Among the South Sea Islands* (New York: Scribner's, 1914), 119.

15   Born in the Scottish Highlands: "The Loss of the Florence Nightingale," *Glasgow Herald*, February 13, 1872.

16   The annals of blackbirding: "120 Died on the Black Doctor's Slaughter Ship," *Truth* (Sydney), January 22, 1950; "The Carl Outrages," *Argus* (Melbourne, Victoria), December 21, 1872.

16   John Cameron likewise used violence: John Cameron, transcribed by Andrew Farrell, *John Cameron's Odyssey* (New York: Macmillan Company, 1928), 174; J. A. Bennett, "Immigration, 'Blackbirding,' Labour Recruiting? The Hawaiian Experience 1877–1887," *The Journal of Pacific History* 11, no. 1 (1976): 3–27.

16   In a sinister inversion: "A Sale of Souls," *San Francisco Examiner*, October 15, 1892.

17   John attributed: Cameron, *Odyssey*, 176

17   At lush, mountainous Pentecost Island: Cameron, *Odyssey*, 182–84.

17   After his lucrative career: "Flew a Black Flag," *Evening Bulletin* (Honolulu), April 12, 1894.

18   By then in his midthirties: Cameron, *Odyssey*, 237; "Death of Miss Barbour," *Honolulu Advertiser*, March 1, 1886; "Local News," *Daily Honolulu Press*, March 1, 1886; "Funeral of Miss Barbour," *Daily Honolulu Press*, March 2, 1886.

18   Everyone on board: "Wreck of the Steamer Planter," *Honolulu Advertiser*, February 1, 1886.

18   The *Planter* broke in two: "Niihau," *Evening Bulletin* (Honolulu), May 10, 1886; "Local News," *Daily Honolulu Press*, April 5, 1886.

19   Kingdom leaders: "The Paradise of the Pacific," *Pacific Commercial Advertiser*, September 30, 1887; Robert C. Schmitt, *Historical Statistics of Hawaii* (Honolulu: University Press of Hawaii, 1977), 273; James Mak, "Creating 'Paradise of the Pacific': How Tourism Began in Hawaii," *UHERO Working Paper No. 2015-1*, Economic Research Organization at the University of Hawai'i, February 2015, 6, 15.

19   The crash of the *Planter*: Mak, "Creating 'Paradise of the Pacific,'" 14; Thomas G. Thrum, ed., *Hawaiian Almanac and Annual for 1891* (Honolulu: Press Publishing, 1890), 88.

19   Once Florence arrived: Leigh H. Irvine, ed., *A History of the New California Its Resources and People*, vol. 2 (New York: Lewis Publishing: 1903), 866–67.

20   In addition to Florence's condition: "Death of Miss Barbour," *Honolulu Advertiser*, March 1, 1886; Liliuokalani, *Hawaii's Story by Hawaii's Queen* (Boston: C. J. Peters & Son, 1898), 61.

20   Dr. Martin tended to Florence: Irvine, *A History*, 866–67; "Local News," *Daily Honolulu Press*, March 1, 1886; "Funeral of Miss Barbour," *Daily Honolulu Press*, March 2, 1886; Rev. Louis F. Benson, *The Best Church Hymns* (Philadelphia: Westminster Press, 1899), 6.

21   Showing an adroitness: Cameron, *Odyssey*, 237; "Flew a Black Flag," *Evening Bulletin* (Honolulu), April 12, 1894; "Rumors," *Evening Bulletin* (Honolulu), August 18, 1886; "Local and General News," *Evening Bulletin* (Honolulu), July 8, 1886; "Local News," *Daily Honolulu Press*, April 14, 1886; "The Past Year," *Evening Bulletin* (Honolulu), January 5, 1886.

21   John's partner: Cameron, *Odyssey*, 237; "Flew a Black Flag," *Evening Bulletin* (Honolulu), April 12, 1894; "Activity of the Volcano," *Honolulu Advertiser*, October 1, 1886; "Shipping Notes," *Honolulu Advertiser*, October 29, 1886;

21   With another venture: Cameron, *Odyssey*, 240.

Chapter 3: The Skies Darken

23   As the *Wandering Minstrel* continued: Storrs L. Olson, "History and Ornithological Journals of the Tanager Expedition of 1923 to the Northwest Hawaiian Islands, Johnston and Wake Islands," *Atoll Research Bulletin* 433 (1996): 85; Frederick D. Walker, *Log of the Kaalokai* (Honolulu: Hawaiian Gazette Co., 1909), 15.

23   According to the superstitions: "The Voyage of the Hawaiian Brig Allie Rowe," *Honolulu Advertiser*, August 28, 1885.

24   Some shark fishermen: Alfred Wilks Drayson, "Sharks," in *Routledge's Every*

Boy's Annual, ed., George Routledge (London: Routledge and Sons, 1877), 522; "Sharks Swim with Their Heads Off," *Daily Pacific Commercial Advertiser*, February 4, 1885; Tobey Pearl, "Twice Bitten: The Hidden History of Sharks in Massachusetts," *On the Water*, August 2020, 82–85.

24   John Cameron believed: "Sons of Captain Walker Refute Wandering Minstrel Story," *Honolulu Advertiser*, January 30, 1922.

24   Conditions at sea: Walker, *Log*, 54; Robert Louis Stevenson and Lloyd Osbourne, *The Wrecker* (London: Cassell, 1892), 205.

25   The few people of the era: George C. Munro, "Nature Studies," *Honolulu Star Bulletin*, May 2, 1931.

25   The *Minstrel* anchored: John Cameron, transcribed by Andrew Farrell, *John Cameron's Odyssey* (New York: Macmillan Company, 1928), 275.

26   For three days and nights: Walker, *Log*, 55.

26   The Walkers could not help thinking: Clarice B. Taylor, "Hawaii's Old Gods Live," *Honolulu Star-Bulletin*, October 26, 1940; "Terrible Death," *Honolulu Advertiser*, July 31, 1885.

26   Down below in the hold: Cameron, *Odyssey*, 279–80.

27   John and Mouwen made it: "The Marloo Fatality," *West Australian*, January 9, 1897.

27   On the rapidly foundering *Minstrel*: "Sons of Captain Walker Refute Wandering Minstrel Story," *Honolulu Advertiser*, January 30, 1922.

27   The dog appeared: Elizabeth Walker, "Wrecked on Midway Island in 1888," *Paradise of the Pacific*, November 1936, 27; Cameron, *Odyssey*, 280.

Chapter 4: The Edge of the World

29   The castaways awoke: "Ocean and Midway Island," *The Friend* (Honolulu), February 1, 1871; William Henry Rosser, *North Pacific Pilot: The Seaman's Guide to the Islands of the North Pacific* (London: James Irmay and Sons, 1870), 62; "Death of George Hanker," *Hilo Daily Tribune*, December 13, 1901.

29   Steering the small lifeboats: John Cameron, transcribed by Andrew Farrell, *John Cameron's Odyssey* (New York: The Macmillan Company, 1928), 281–84;

Frederick D. Walker, *Log of the Kaalokai* (Honolulu: Hawaiian Gazette Co., 1909), 55.

31   Older sailors among the crew: "Death of George Hanker," *Hilo Daily Tribune*, December 13, 1901.

31   Despite all: John Seelye, introduction to *The Swiss Family Robinson*, by Johann D. Wyss (New York: Penguin, 2007), vii–xxii; Karen Sánchez-Eppler, "Castaways: The Swiss Family Robinson, Child Bookmakers, and the Possibilities of Literary Flotsam," in *Oxford Handbook of Children's Literature*, ed. Julia L. Mickenberg and Lynne Vallone (Oxford: Oxford University Press, 2011), 433–54.

32   The first person documented: William Reynolds to H. K. Thatcher, September 30, 1867, in *Senate Documents, Otherwise Publ. as Public Documents and Executive Documents 14th Congress, 1st Session-48th Congress, 2nd Session and Special Session* 7 (Washington, DC: Government Printing Office, 1868), 7.

33   That doomed mission: "Ocean and Midway Island," *The Friend* (Honolulu), February 1, 1871.

33   Naval officials: George Read, *The Last Cruise of the Saginaw* (Boston: Houghton Mifflin, 1912), 6–10.

34   The *Saginaw*'s mission: Read, *Last Cruise*, 8–14; "The Story of the 'Saginaw,'" *Harper's Weekly*, March 4, 1871, 197–98; Montgomery Sicard to Thomas Turner, July 21, 1870, in *Annual Report of the Secretary of the Navy on the Operations of the Department for the Year 1871* (Washington, DC: Government Printing Office, 1871), 226–27.

35   Naval personnel: Read, *Last Cruise*, 14; Montgomery Sicard to Henry A. Pierce, November 1870, in *Papers Relating to the Foreign Relations of the United States* (Washington, DC: Government Printing Press, 1871), 546.

35   Approaching Ocean Island: Read, *Last Cruise*, 21, 67–68.

36   The crew on the lifeboat: Hanson W. Baldwin, "The Narrative of William Halford," *Proceedings* 61, no. 391 (September 1935): 1287–97; "Reception at Iolani Palace," repr. in *Annual Report of the Secretary of the Navy*, 209; Montgomery Sicard to John A. Winslow, February 9, 1871, in *Annual Report*

of the Secretary of the Navy, 219; "Hawaii vs. China," *Hawaiian Gazette*, January 18, 1871.

37   Though the wreck had occurred: "Naval Notes," *Honolulu Advertiser*, February 11, 1889; "Wandering Minstrel," *Hawaiian Star*, April 11, 1895.

37   Taking into account: "Shipwreck of Ship Henry James," *Thames Star*, May 26, 1888; Rosser, *North Pacific Pilot*, 64; Mark J. Rauzon, *Isles of Refuge: Wildlife and History of the Northwestern Hawaiian Islands* (Honolulu: University of Hawaii Press, 2001), 34.

38   John Cameron noted: Cameron, *Odyssey*, 290.

38   The castaways had to find: Bennet Woodcroft, *Chronological and Descriptive Index of Patents Applied for and Patents Granted* (London: George Edward Eyre and William Spottiswoode, 1875), 660.

38   While they experimented: Cameron, *Odyssey*, 286–87.

38   Frederick felt ill: Walker, *Log*, 41.

39   On Sand Island: Cameron, *Odyssey*, 285–93.

39   After tending to his wound: Robert Louis Stevenson and Lloyd Osbourne, *The Wrecker* (London: Cassell, 1892), 377.

Chapter 5: State, War, Navy

41   The fate of Midway: Joseph West Moore, *Picturesque Washington: Pen and Pencil Sketches* (Providence, RI: J. A. & R. A. Reid, 1888), 167.

41   Beyond a grand entrance: Benjamin Perley Poore, *Perley's Reminiscences of Sixty Years in the National Metropolis*, vol. 2 (Philadelphia: Hubbard Brothers, 1886), 360.

42   Bayard's State Department: "Ex-Governor Curtin Mad," *Daily Republican* (Wilmington, Delaware), October 6, 1885.

42   To make his case: "Naval Notes," *Honolulu Advertiser*, February 11, 1889; George Read, *The Last Cruise of the Saginaw* (Boston: Houghton Mifflin, 1912), 101.

43   When the time came: "Secretary Bayard as a Psychologist," *Seattle Post-Intelligencer*, September 5, 1885; Sir Willoughby Maycock, *With Mr. Chamberlain in the United States and Canada, 1887–1888* (London: Chatto & Windus, 1914), 62.

43   The full briefing: Thomas Bayard to George H. Pendleton, January 17, 1888, in *Papers Relating to the Foreign Relations of the United States* (Washington, DC: Government Printing Office, 1889), 594–608; Charles Callan Tansill, *Diplomatic Relations Between the United States and Hawaii, 1885–1889* (New York: Fordham University Press, 1940), 24.

44   But certain details: "The Legislative Assembly," *Hawaiian Gazette*, November 8, 1887.

44   Bayard doubted: Ralph Kuykendall, *The Hawaiian Kingdom, Volume 3, 1874–1893: The Kalakaua Dynasty* (Honolulu: University of Hawaii Press, 1967), 415–16.

45   At the center of this plot: Albin Kowalewski, *Asian and Pacific Islander Americans in Congress, 1900–2017* (Washington, DC: Government Printing Office, 2017), x, 102; Kuykendall, *Hawaiian Kingdom*, 416–18; Liliuokalani, *Hawaii's Story by Hawaii's Queen* (Boston: C. J. Peters & Son, 1898), 193.

45   Those in the Hawaiian government: "The Canadian Abroad," *Evening Bulletin* (Honolulu), March 9, 1888.

46   That brought Bayard back: *Journal of the Executive Proceedings of the Senate of the United States*, vol. 26 (Washington, DC: Government Printing Office, 1901), 92; Poore, *Perley's Reminiscences*, 489.

46   As his ongoing series: "Secretary Bayard as a Psychologist," *Seattle Post-Intelligencer*, September 5, 1885.

46   All the developments: Frances P. Thomas, *Career of John Grimes Walker, U.S.N., 1835–1907* (Boston: privately printed, 1959), 61.

Chapter 6: The Stranger

48   The *Wandering Minstrel* castaways: Frederick D. Walker, *Log of the Kaalokai* (Honolulu: Hawaiian Gazette Co., 1909), 59.

48   The hut: "A Sale of Souls," *San Francisco Examiner*, October 15, 1892; John Cameron, transcribed by Andrew Farrell, *John Cameron's Odyssey* (New York: Macmillan Company, 1928), 252.

49   As for the dynamics: Cameron, *Odyssey*, 300.

50   Frederick, for his part: Lyle S. Shelmidine, "The Early History of Midway Islands," *American Neptune*, July 1948, 188.

50   Fish, as it turned out: Walker, *Log*, 59; Cameron, *Odyssey*, 284–87; Elizabeth Walker, "Wrecked on Midway Island in 1888," *Paradise of the Pacific*, November 1936, 27.

51   While the castaways organized: "Naval News and Views," *Norfolk Landmark*, January 29, 1888.

52   A Massachusetts native: *The United Service: A Monthly Review of Military and Naval Affairs*, vol. 4 (Philadelphia: L. R. Hamersly, 1881), 17.

52   By the time they received: "From the Juniata," *Evening Bulletin* (Honolulu), April 6, 1888; "Island Locals," *Hawaiian Gazette*, September 27, 1887; "Naval Notes," *Honolulu Advertiser*, February 11, 1889.

52   Though they fell short: "Death of Prince Edward," *Hawaiian Gazette*, September 27, 1887; "Foreign Office," *Evening Bulletin* (Honolulu), September 28, 1887; "Local and General," *Honolulu Advertiser*, September 29, 1887; "The Legislative Assembly," *Hawaiian Gazette*, November 8, 1887; "Local and General," *Honolulu Advertiser*, December 30, 1887.

53   *Juniata* was an ideal vessel: United States Hydrographic Office, Navy Department, *Notice to Mariners* 8 (February 26, 1887): 54–55; George M. Robeson, "Proceedings of the Juniata in the Search for the Polaris," in *Report of the Secretary of the Navy* (Washington, DC: Government Printing Office, 1873), 215–16; "Home," *Lloyd's List*, February 3, 1876.

53   The *Juniata* had left: "The Voyage of the U.S.S. 'Juniata,'" *Evening Bulletin* (Honolulu), May 7, 1888; "From the Juniata," Evening Bulletin (*Honolulu*), April 6, 1888.

54   Even after leaving: "The Voyage of the U.S.S. 'Juniata,'" *Evening Bulletin* (Honolulu), May 7, 1888; "From the Juniata," Evening Bulletin (*Honolulu*), April 6, 1888.

55   The Walkers were watching: George Read, *The Last Cruise of the Saginaw* (Boston: Houghton Mifflin, 1912), 78–79; Deirdre O'Regan, Hans Van

Tilburg, and Kelly Gleason, "Whaling Shipwrecks in the Northwestern Hawaiian Islands: The 2008 Maritime Heritage Archeological Expedition," *Sea History* 125 (Winter 2008/2009): 18.

56 All the while: Walker, "Wrecked," 27.

56 The castaways could be divided: "Sons of Captain Walker Refute Wandering Minstrel Story," *Honolulu Advertiser*, January 30, 1922.

56 The escape plan: Edwin H. Bryan, Jr., ed., *Northwestern Hawaiian Islands* (Honolulu: U.S. Fish and Wildlife Service, 1978), 24; "Shipwreck of Ship Henry James," *Thames Star*, May 26, 1888.

57 However, there were differences: Cameron, *Odyssey*, 292.

57 The knowledge gained: Walker, *Log*, 59.

58 Frank Lord, the cook: "Hawaii Notes," *Daily Honolulu Press*, February 19, 1881; "From South Seas," *Hawaiian Gazette*, December 4, 1896.

Chapter 7: Shipwreck of Faith

59 The morning of March 15: "The Wandering Minstrel," *San Francisco Chronicle*, July 1, 1889.

59 "Soon," wrote first mate: John Cameron, transcribed by Andrew Farrell, *John Cameron's Odyssey* (New York: Macmillan Company, 1928), 301.

60 Meanwhile, in Washington, D.C.: "Naval News and Views," *Norfolk Virginian*, March 21, 1888; "Current Comment," *Burlington Democrat* (Burlington, Kansas), April 27, 1888.

61 With shipping so crucial: "The Orient," *San Francisco Examiner*, April 15, 1888; "Local and General News," *Evening Bulletin* (Honolulu), April 5, 1888.

61 Unbeknownst to the castaways: Lloyd's Register of British and Foreign Shipping, *Universal Register*, April 1, 1887–March 31, 1888, listing HAA-HAL.

61 The son of a lieutenant: "Death of Captain H. Stop," *Liverpool Weekly Mercury*, March 4, 1914; "Giant Three Master," *Tacoma Daily News*, December 20, 1897.

62 *Halewood*'s latest passage: "From the Front," *San Francisco Examiner*, March 23, 1888.

62   Crossing the equator: "Captain Henry Stap," *Journal of Commerce*, March 7, 1914.

62   But a hurricane: "Water Front Notes," *San Francisco Examiner*, April 26, 1888.

63   That fact attracted: "From the Front," *San Francisco Examiner*, March 23, 1888.

63   The remaining castaways: "Harriet Beecher Stowe," *St. Louis Globe-Democrat*, September 16, 1888.

64   Though the castaways: 2 Cor. 11:25 (King James Version).

64   One such storied voyage: Acts 27:10 (King James Version).

65   The narrative contained: George Burder, *Sea Sermons: Or, Twelve Short and Plain Discourses for the Use of Seamen* (London: B. Bensley, 1822), 95; Rev. Richard Prichard, trans. Rev. John Hugh Morgan, *The Theological Instructor: Or, A Scriptural View of the Evidences, Doctrines, Morals, and Institutions of the Christian Religion* (London: George Philip & Son, 1875), 145; Frederick D. Walker, *Log of the Kaalokai* (Honolulu: Hawaiian Gazette Co., 1909), 24.

65   The stolen boat: Elizabeth Walker, "Wrecked on Midway Island in 1888," *Paradise of the Pacific*, November 1936, 27–28.

66   Elizabeth became: Walker, "Wrecked," 27–28.

66   July 4 brought: Walker, *Log*, 46.

Chapter 8: Hans's Story, Part I

69   The path that led Hans Jorgensen: "Notes," *Daily Honolulu Press*, April 28, 1886; "The General Siegel," *Honolulu Advertiser*, December 7, 1887; "Death of Henry F. Glade Occurs in Fatherland," *Honolulu Advertiser*, January 29, 1902; "A Brilliant Ball," *Evening Bulletin* (Honolulu), October 22, 1889.

69   Proving he could: "The Ocean Cable Enterprise," *Hawaiian Gazette*, July 23, 1889; "The Blackbird Cruise," *San Francisco Examiner*, October 16, 1892.

70   Glade was unconvinced: "Local and General," *Honolulu Advertiser*, February 4, 1885.

70   Inconsistent oversight: "Local and General News," *Evening Bulletin* (Honolulu), August 28, 1886; *Evening Bulletin* (Honolulu), September 14, 1885; "Local and General News," *Daily Bulletin* (Honolulu), September 15, 1885; "Local & General News," *Daily Bulletin* (Honolulu), November 26, 1885.

71   On September 1, 1886: "The General Siegel," *Honolulu Advertiser*, December 7, 1887.

71   By this point: "The General Siegel," *Honolulu Advertiser*, December 7, 1887.

71   Sanders assessed: Edvart Olsen, "The Midway Tragedy," *Mid-Pacific Magazine*, January 1918, 78.

71   Meanwhile, approximately ten days: F. J. Talfourd Charter, "The Relations of the State with Fishermen and Fisheries," in *The Fisheries Exhibition Literature*, vol. 9 (London: William Clowes and Son, 1884), 25354; J. C. Wilcocks, "Improved Fishery Harbour Accommodation for Great Britain and Ireland," in *The Fisheries Exhibition*, 34.

72   The spare scow: Olsen, "Midway Tragedy," 78.

73   The *Dunnottar Castle* wreck: "Local and General News," *Daily Bulletin* (Honolulu), August 25, 1886; "The Waialeale," *Daily Bulletin* (Honolulu), August 27, 1886; "Succor," *Honolulu Advertiser*, September 15, 1886.

73   King Kalākaua came to the wharf: "Personal Items," *Rural New-Yorker*, February 26, 1881; "Succor," *Honolulu Advertiser*, September 15, 1886.

74   But neither Wodehouse: "Return of the S. S. Waialeale from Ocean Island," *Daily Bulletin* (Honolulu), September 29, 1886; "The Ke Au Hou," *Honolulu Advertiser*, June 17, 1887.

74   To the *General Siegel* castaways: "The Shipwreck," *Daily Herald*, September 15, 1886.

75   Sanders was right: "Where Is the General Siegel?," *Daily Bulletin*, February 17, 1887.

75   A lost or missing ship: "Work for the Kaimiloa," *Pacific Commercial Advertiser*, April 27, 1887; "Where is the General Siegel?," *Daily Bulletin*, February 17, 1887; "H. H. M. S. Kaimiloa," *Hawaiian Gazette*, February 1, 1887.

75   None of this was known: Andrew Farrell, "Island Wrecks VII," *Pacific Marine Review*, November 1920, 57.

76   "Arrived September 1886": Lyle S. Shelmidine, "The Early History of Midway Islands," *American Neptune*, July 1948, 186.

77   Hans backed away: "The General Siegel," *Honolulu Advertiser*, December 7, 1887.

Chapter 9: Hans's Story, Part II

78   While urgency mounted: "A Search Expedition," *Daily Herald* (Honolulu), May 2, 1887.

78   On Sand Island: Many details of the wreck of the *General Siegel* in this section are drawn from "The General Siegel," *Honolulu Advertiser*, December 7, 1887; Edvart Olsen, "The Midway Tragedy," *Mid-Pacific Magazine*, January 1918, 77–81; John Cameron, transcribed by Andrew Farrell, *John Cameron's Odyssey* (New York: Macmillan Company, 1928), 250–60.

79   Sanders, Brown, and Hans: Robert Louis Stevenson and Lloyd Osbourne, *The Wrecker* (London: Cassell, 1892), 394.

81   All the while: "The Ke Au Hou," *Honolulu Advertiser*, June 17, 1887.

84   Hans was alone: Stevenson and Osbourne, *Wrecker*, 214.

Chapter 10: The Monarch of All

85   Hans had transformed: John Cameron, transcribed by Andrew Farrell, *John Cameron's Odyssey* (New York: Macmillan Company, 1928), 260–64; Robert Louis Stevenson and Lloyd Osbourne, *The Wrecker* (London: Cassell, 1892), 384.

86   Seven months into his isolation: Lyle S. Shelmidine, "The Early History of Midway Islands," *American Neptune*, July 1948, 187.

86   As for Hans's quick affinity: "From South Seas," *Pacific Commercial Advertiser*, December 4, 1896; Cameron, *Odyssey*, 251.

86   George, in fact, continued: "Death of George Hanker," *Hilo Daily Tribune*, December 13, 1901; Cameron, *Odyssey*, 282.

87   With the loss of George's companionship: "Survivors Now Here," *Honolulu Advertiser*, August 15, 1900.

87   But Hans was not the only one: Edvart Olsen, "The Midway Tragedy," *The Mid-Pacific Magazine*, January 1918, 80–81.

88   The version of events: "Local & General News," *Daily Bulletin* (Honolulu), December 9, 1887.

89 Frederick was by nature: "A Honolulu Man Knows Where Much Silver Has Been Buried," *Honolulu Advertiser*, July 6, 1903.

89 The *Minstrel*'s shipwreck: "Sons of Captain Walker Refute Wandering Minstrel Story," *Honolulu Advertiser*, January 30, 1922; "Local & General News," *Daily Bulletin* (Honolulu), December 9, 1887; Olsen, "Midway Tragedy," 78–79 (referring to Sand Island for what should have been Green Island).

90 He had to protect Hans: Cameron, *Odyssey*, 285.

90 It remains uncertain: Elizabeth Walker, "Wrecked on Midway Island in 1888," *Paradise of the Pacific*, November 1936, 27–28; "Sons of Captain Walker Refute Wandering Minstrel Story," *Honolulu Advertiser*, January 30, 1922.

90 Over time: Cameron, *Odyssey*, 260.

91 George Hanker had special reason: "Death of George Hanker," *Hilo Daily Tribune*, December 13, 1901.

91 With Hans's true nature exposed: William Cowper, *The Poems of William Cowper*, vol. 2 (Chiswick, England: C. Whittingham, 1822), 196.

92 But if an uninhabited island: "The Wandering Minstrel—50 Years After," *Honolulu Star-Bulletin*, June 25, 1938.

92 There had been something else: Olsen, "Midway Tragedy," 78; "Local & General News," *Daily Bulletin* (Honolulu), September 21, 1886.

92 At one point after this: Frederick Chamberlin, *Around the World in Ninety Days* (Boston: Clark Publishing, 1906), 74; "Exiled for His Conscience," *Fort Worth Star Telegram*, May 9, 1912.

Chapter 11: Tempests

94 Mid-August brought: John Cameron, transcribed by Andrew Farrell, *John Cameron's Odyssey* (New York: The Macmillan Company, 1928), 295.

94 That night: Frederick D. Walker, *Log of the Kaalokai* (Honolulu: Hawaiian Gazette Co., 1909), 61; "On a Mid-Ocean Island," *San Francisco Call*, February 26, 1895.

95 It was a hurricane: Elizabeth Walker, "Wrecked on Midway Island in 1888," *Paradise of the Pacific*, November 1936, 27; Cameron, *Odyssey*, 296.

95 In fact, Hans was in severe danger: Cameron, *Odyssey*, 296–97.

95   Meanwhile, out on the open Pacific: "Last night," *Auckland Star*, October 27, 1891; "The Shaw, Savill and Albion Company," *Auckland Star*, December 22, 1890.

96   Scenarios involving intentional wrecks: Jeremy Borrelli, "Initial Assessment of Lead Artefacts Used for Hull Repair and Maintenance on North Carolina Shipwreck 31CR314, *Queen Anne's Revenge*," *International Journal of Nautical Archaeology* 49, no. 2 (June 2020): 357–70.

96   In the fraud scheme: "A Bark Set on Fire," *San Francisco Chronicle*, April 1, 1888; "Who Burned Her?" *San Francisco Examiner*, April 1, 1888.

97   Yet another version: Prison and Correctional Records, Folsom State Prison, Folsom, California, 1888, 127–28.

98   An Episcopal bishop: William Ingraham Kip, *The Early Days of My Episcopate* (New York: Thomas Whittaker, 1892), 70.

98   In the middle of the next night: "Ship on Fire," *San Francisco Examiner*, March 10, 1888.

99   Reviewing the circumstances: "A Bark Set on Fire," *San Francisco Chronicle*, April 1, 1888.

99   These discoveries: "The Evil Eye," *San Francisco Examiner*, May 9, 1888.

99   Both Caskey and Bachman: "Isaiah Lees Dies, But Fame as Detective Will Long Live," *San Francisco Call*, December 22, 1902; "Lees Grabs Some More Greatness," *San Francisco Call*, April 29, 1898; "Sovereigns, Brass Bands and Detectives, and Now Richard Roe and John Doe Are Gone," *San Francisco Examiner*, September 2, 1899.

100   The critique that Lees: "Ship on Fire," *San Francisco Examiner*, March 10, 1888; "The Evil Eye," *San Francisco Examiner*, May 9, 1888; "Some of Them Are Mad," *San Francisco Examiner*, December 18, 1895.

100   However, Captain Crosby: "Bay Gleanings," *Daily Bee* (Sacramento, CA), August 25, 1888.

101   Back on Sand Island: Cameron, *Odyssey*, 297–98; "On a Mid-Ocean Island," *San Francisco Call*, February 26, 1895.

101   John and Hans moved: "Sons of Captain Walker Refute Wandering Minstrel

Story," *Honolulu Advertiser*, January 30, 1922; George C. Munro, "Nature Studies," *Honolulu Star Bulletin*, May 2, 1931.

102   Before joining the *Siegel*: Melissa Macauley, *Distant Shores: Colonial Encounters on China's Maritime Frontier* (Princeton, NJ: Princeton University Press, 2021), 162 (on sugar cargo as pretext for opium).

102   The unusual circumstances: "Strange Suspicions," Daily Pacific Commercial Advertiser, June 12, 1889.

103   *The Wrecker*: Robert Louis Stevenson and Lloyd Osbourne, *The Wrecker* (London: Cassell, 1892), 233.

104   Midway represented: Walker, *Log*, 396.

104   The Walkers did not possess: Cameron, *Odyssey*, 449.

104   The Walker boys tended to stay: "Well Known Character Dead," *Hawaii Herald*, December 12, 1901.

105   Still, Frederick had to face: "Sons of Captain Walker Refute Wandering Minstrel Story," *Honolulu Advertiser*, January 30, 1922.

105   Colder weather set in quickly: Cameron, *Odyssey*, 299–303.

106   They could not know: US Navy Department, Bureau of Construction and Repair, *Annual Report of the Chief of the Bureau of Construction and Repair to the Secretary of the Navy for the Fiscal Year Ending June 30, 1889* (Washington, DC: Government Printing Office, 1889), 15.

107   The looming showdown: Cameron, *Odyssey*, 302–3; "Sons of Captain Walker Refute Wandering Minstrel Story," *Honolulu Advertiser*, January 30, 1922; Walker, "Wrecked," 27.

108   On October 13, 1888: "Wandering Minstrel," *Honolulu Advertiser*, April 8, 1889; "The Wandering Minstrel," *San Francisco Chronicle*, July 1, 1889; Cameron, *Odyssey*, 304.

108   Watching Hans cast off: David J. Skal, *Something in the Blood: The Untold Story of Bram Stoker, the Man Who Wrote Dracula* (New York: Liveright Publishing Company, 2016), 282–83.

108   After John, Hans, and Mouwen: "From South Seas," *Daily Commercial Pacific Advertiser*, December 4, 1896.

Chapter 12: *Lights: Fall*

110   John and Hans's theft: Frederick D. Walker, *Log of the Kaalokai* (Honolulu: Hawaiian Gazette Co., 1909), 56; "The Wandering Minstrel," *San Francisco Chronicle*, July 1, 1889; Elizabeth Walker, "Wrecked on Midway Island in 1888," *Paradise of the Pacific*, November 1936, 28.

110   In the day-to-day repetition: Robert Louis Stevenson and Lloyd Osbourne, *The Wrecker* (London: Cassell, 1892), 205, 375.

110   In the weeks following: George C. Munro, "Nature Studies," *Honolulu Star Bulletin*, May 2, 1931.

111   One evening: "Survivors Now Here," *Honolulu Advertiser*, August 15, 1900; S. B. Luce, rev. by Aaron Ward, *Text-Book of Seamanship* (New York: D. Van Nostrand, 1884), 360–74.

111   Elizabeth described: Walker, "Wrecked," 28; "The Wandering Minstrel," *San Francisco Chronicle*, July 1, 1889.

111   Those lights the castaways saw: "The SS City of Peking," *Daily Alta* (California), January 28, 1875.

112   At *Peking*'s lowest level: "An Ocean Palace: The Launch of the City of Peking," *New York Times*, March 19, 1874; Marcy C. Greenfield, "Benevolent Desires and Dark Dominations: The Pacific Mail Steamship Company's SS City of Peking and the United States in the Pacific 1874–1910," *Southern California Quarterly* 94, no. 4 (Winter 2012): 426.

112   At the helm of *Peking*: A. C. Oudemans, *The Great Sea-serpent: An Historical and Critical Treatise* (Leiden, Netherlands: privately printed, 1892), 48; E. W. Wright, ed., *Lewis & Dryden's Marine History of the Pacific Northwest* (Portland, OR: Lewis & Dryden Printing, 1895), 235; "Liner Korea Brings Over a Thousand Passengers from the Far East," *San Francisco Call*, April 21, 1905; In Re Pacific Mail S. S. Co., No. 1,035 (Circuit Court of Appeals, May 2, 1904), 81.

113   The *Peking* carried cargo: "Trade with the Orient," *San Francisco Examiner*, October 28, 1888.

113   On its most recent arrival: David M. Pletcher, *The Diplomacy of Involvement: American Economic Expansion* (Columbia, MO: University of Missouri Press, 2001), 149; Mary Roberts Coolidge, *Chinese Immigration* (New York: Henry Holt, 1909), 183.

114   With *Peking*'s status: "The Arabic in Port," *San Francisco Examiner*, October 27, 1888.

114   Even the Chinese crew members: "Trade with the Orient," *San Francisco Examiner*, October 28, 1888; In Re Pacific Mail S. S. Co., No. 1,035 (Circuit Court of Appeals, May 2, 1904), 81.

114   A sense of how different: Rudyard Kipling, *From Sea to Sea*, vol. 1 (New York: Doubleday, 1899), 424–25.

114   Captain Seabury also wanted: Amory H. Bradford, "The Japan Deputation at Honolulu," *The Independent* (New York), April 9, 1896.

115   Seabury knew: "Liner Korea Brings Over a Thousand Passengers from the Far East," *San Francisco Call*, April 21, 1905; "The Exclusion Act," *Los Angeles Times*, October 28, 1888; "Survivors Now Here," *Honolulu Advertiser*, August 15, 1900.

115   With the excitement: George Read, *The Last Cruise of the Saginaw* (Boston: Houghton Mifflin, 1912), 87.

Chapter 13: Master of Misdirection

117   Nobody had heard: John Cameron, transcribed by Andrew Farrell, *John Cameron's Odyssey* (New York: The Macmillan Company, 1928), 302–5.

117   By the middle: Cameron, *Odyssey*, 305; George Read, *The Last Cruise of the Saginaw* (Boston: Houghton Mifflin, 1912), 61.

118   The fact was: "Wandering Minstrel," *Honolulu Advertiser*, April 8, 1889; Frederick D. Walker, *Log of the Kaalokai* (Honolulu: Hawaiian Gazette Co., 1909), 59–60.

118   Meanwhile, dark clouds: Cameron, *Odyssey*, 305–7.

118   Rain fell: Cameron, *Odyssey*, 307–10.

118   In case they ran out: Cameron, *Odyssey*, 302, 324.

119   After two weeks: Cameron, *Odyssey*, 311–23.

119   At Mili: A contemporary account for the mutineers on Mili can be found at
Cyrus M. Hussey and William Lay, *A Narrative of the Mutiny, on Board the
Ship Globe, of Nantucket, in the Pacific Ocean, Jan. 1824, and the Journal of a
Residence of Two Years on the Mulgrave Islands; with Observations on the Manners
and Customs of the Inhabitants* (New London, CT: 1828).

120   As the latest visitors: Cameron, *Odyssey*, 197, 326–35.

121   Enough time: "One Thing and Another," *Peterhead Sentinel and General
Advertiser for Buchan District*, May 3, 1889.

121   Settling into Jaluit: Cameron, *Odyssey*, 330–31; "The Whereabouts of Captain
Cameron, of the Ebon," *Hawaiian Gazette*, April 20, 1894; *Commercial Relations
of the United States Reports from the Consuls of the United States on the Commerce,
Manufactures, Etc., of Their Consular Districts* (Washington, DC: Government
Printing Office, 1889), 56–58.

122   When asked: Edward H. M. Davis, *Papers Respecting the Reclaration of a British
Protectorate over the Gilbert Islands* (London: Royal Empire Society, 1892),
146; United States Hydrographic Office, Navy Department, *The Hawaiian
Islands and the Islands, Rocks, and Shoals to the Westward* (Washington, DC:
Government Printing Office, 1899), 53.

122   The ship name: "Sons of Captain Walker Refute Wandering Minstrel Story,"
Honolulu Advertiser, January 30, 1922; "Disappearance of a Pacific Island,"
Christian Advocate, March 15, 1900; Philip Marsden, "Deleted Islands," 1843
magazine, *The Economist*, September/October 2015.

123   The first reward: "Information Wanted," *Evening Bulletin* (Honolulu), March
6, 1889; Ralph Kuykendall, *The Hawaiian Kingdom, Volume 3, 1874–1893:
The Kalakaua Dynasty* (Honolulu: University of Hawaii Press, 1967), 421–22;
Nancy J. Morris, introduction to *The Biography of Hon. Robert William Wilcox*,
by Thomas K. Nananaela (Honolulu: privately published, 1993), xi.

124   Eschewing Hawaii: "On a Desert Island," *San Francisco Chronicle*, April 21,
1889; "Our San Francisco Letter," *Honolulu Advertiser*, March 8, 1889.

124   Leaving Hans: "Correspondence," *The Monthly Journal Published by the Brotherhood of Locomotive Engineers*, September 1876, 401.

125   John sat down: "Strange Suspicions," *Daily Pacific Commercial Advertiser*, June 12, 1889; "Our San Francisco Letter," *Honolulu Advertiser*, March 8, 1889; "Capt. Walker Fully Exonerated," *Daily Commercial Pacific Advertiser*, April 4, 1889; "A Derelict Bark," *San Francisco Examiner*, February 15, 1889.

Chapter 14: Wail of Lost Souls: Winter

126   Enough time passed: "Sons of Captain Walker Refute Wandering Minstrel Story," *Honolulu Advertiser*, January 30, 1922; "On a Mid-Ocean Island," *San Francisco Call*, February 26, 1895.

126   Ten months: "Shipwreck of Ship Henry James," *Thames Star*, May 26, 1888.

126   The ever-changing: Diary of A. L. C. Atkinson, reprinted in John Cameron, transcribed by Andrew Farrell, *John Cameron's Odyssey* (New York: Macmillan Company, 1928), 439, n30.

127   Many who spent time: Cameron, *Odyssey*, 439–40, n30; Frederick D. Walker, *Log of the Kaalokai* (Honolulu: Hawaiian Gazette Co., 1909), 54.

127   Edvart Olsen: Edvart Olsen, "The Midway Tragedy," *The Mid-Pacific Magazine*, January 1918, 81.

127   But food now disappeared: Cameron, *Odyssey*, 299; Walker, *Log*, 56; "Hundreds of Seabirds Killed by Strong Storms That Slammed Midway Atoll," Hawaii News Now, January 20, 2022, https://www.hawaiinewsnow.com/2022/01/20 /severe-weather-strikes-papahanaumokuakea-killing-hundreds-seabirds/ (modern example of a hurricane affecting Midway's bird population).

128   With so many birds: Walker, *Log*, 56.

128   Three castaways: Walker, *Log*, 56–59, 62.

128   The combination of drinking: Walker, *Log*, 58; "The Wandering Minstrel—50 Years After," *Honolulu Star-Bulletin*, June 25, 1938; George Read, *The Last Cruise of the Saginaw* (Boston: Houghton Mifflin, 1912), 79–80; Lyle S. Shelmidine, "The Early History of Midway Islands," *American Neptune*, July 1948, 189.

129   As for marine life: Cameron, *Odyssey*, 287.

129   With each source of food: "On a Desert Island," *San Francisco Chronicle*, April 21, 1889; Elizabeth Walker, "Wrecked on Midway Island in 1888," *Paradise of the Pacific*, November 1936, 28.

129   Elizabeth's health: Walker, *Log*, 54–55.

129   He decided to take the boat: Three-day fishing expedition in this chapter primarily drawn from Walker, "Wrecked," 28; "On a Mid-Ocean Island," *San Francisco Call*, February 26, 1895; "Survivors Now Here," *Honolulu Advertiser*, August 15, 1900.

130   While horror stories: James William Buel, *The Living World: A Complete Natural History of the World's Creatures* (Philadelphia: Historical Publishing Company, 1891), 77; Clarice B. Taylor, "Hawaii's Old Gods Live," *Honolulu Star-Bulletin*, October 26, 1940.

133   All the castaways: Walker, "Wrecked," 28.

134   Elizabeth continued: Abby Jane Morrell, *Narrative of a Voyage to the Ethiopic and South Atlantic Ocean, Indian Ocean, Chinese Sea, North and South Pacific Ocean, in the Years 1829, 1830, 1831* (New York: Harper, 1833), 19.

134   Even in her own feeble condition: Walker, "Wrecked," 28.

Chapter 15: Riddles

136   Alexander managed: "Protected Policy Holders," *San Francisco Chronicle*, January 1, 1889; "The South British Company," *San Francisco Examiner*, December 29, 1889.

137   Even though John Cameron's story: "The Case of Charles R. Backman," *Baltimore Underwriter A Monthly Publication Devoted to the Interests of Insurance* 39 (June 5, 1888): 282.

137   Recent reports: "Casualties," *Glasgow Herald*, November 23, 1883; "Local & General News," *Daily Bulletin* (Honolulu), December 2, 1887.

138   Alexander understood: "On a Desert Island," *San Francisco Chronicle*, April 21, 1889; "Strange Suspicions," *Honolulu Advertiser*, June 12, 1889.

138   Alexander kept close ties: Robert K. DeArment, *Man-Hunters of the Old West*, vol. 2 (Norman, OK: University of Oklahoma University Press, 2018), 100.

139   He had served: "Procopio, the Bandit," *San Francisco Chronicle*, February 11, 1872; Charles Howard Shinn, *Graphic Description of Pacific Coast Outlaws* (San Francisco: R. R. Patterson, circa 1887), 22.

139   Wrapping up his career: "Ex-Sheriff Morse," *Oakland Tribune*, July 20, 1878; "Black Bart," *San Francisco Examiner*, September 16, 1883; "Black Bart," *San Francisco Examiner*, September 14, 1883.

140   In what he called: John Boessenecker, *Lawman: The Life and Times of Harry Morse, 1835–1912* (Norman, OK: University of Oklahoma Press, 1998), 238–240; "Smugglers at Work," *Daily Alta* (California), May 24, 1887.

140   A month after: "The Long-Lost Rubio," *San Francisco Examiner*, October 28, 1888; "Agnew and His Books," *Oakland Tribune*, April 29, 1890; "Jerome Deasy in Trouble," *Oakland Tribune*, October 19, 1890.

141   Because he was viewed: "Sovereigns, Brass Bands and Detectives, and Now Richard Roe and John Doe Are Gone," *San Francisco Examiner*, September 2, 1899.

142   Any plan by John: "Harkins Convicted," *San Francisco Examiner*, November 12, 1887; "The Pacific Slope," *San Francisco Chronicle*, February 16, 1888.

142   Still, it was harder for smugglers: "An Opium-Filled Bustle," *Seattle Post Intelligencer*, February 20, 1889.

142   For those familiar with Tacoma: Rudyard Kipling, *From Sea to Sea*, vol. 1 (New York: Doubleday, 1899), 115.

143   Around the same time: "Opium Seized at Tacoma," *Seattle Post Intelligencer*, February 26, 1889; "Opium Smuggling," *Spokesman Review*, February 28, 1889.

143   Inquiries in Tacoma: "On a Desert Island," *San Francisco Chronicle*, April 21, 1889; "Strange Suspicions," *Daily Pacific Commercial Advertiser*, June 12, 1889; "Frank Agnew Released," *San Francisco Examiner*, May 7, 1890; "The Mystery Solved," *Daily Pacific Commercial Advertiser*, June 13, 1889.

143   As when he had quickly decamped: John Cameron, transcribed by Andrew

Farrell, *John Cameron's Odyssey* (New York: Macmillan Company, 1928), 332; James A. Michener and A. Grove Day, *Rascals in Paradise* (London: Secker & Warburg, 1957), 224–25. For a biography of Bully Hayes, see Joan Druett, *Notorious Captain Hayes: The Remarkable True Story of William 'Bully' Hayes, The Pirate of the Pacific* (New York: HarperCollins, 2016).

144   Meanwhile, a U.S. naval screw gunboat: "Henry Clay Taylor," in *Dictionary of American Naval Fighting Ships*, ed. James L. Mooney, vol. 7 (Washington, DC: Naval Historical Center, 1981), 69.

144   Part of the Asiatic Squadron: "Death Takes Noted Banker," *Los Angeles Times*, March 14, 1921.

145   Ensign Drake boarded: Cameron, *Odyssey*, 346; Israel Smith Clare, *Naval History of the United States* (New York: Union Book Company, 1899), 289.

145   Ensign Drake felt: *Papers Relating to the Foreign Relations of the United States, Transmitted to Congress, December 5, 1892* (Washington, DC: Government Printing Office, 1893), 452.

146   But that would come after: "Americans in Peril," *San Francisco Chronicle*, January 1, 1891; *Papers Relating to the Foreign Relations of the United States, Transmitted to Congress, December 5, 1892* (Washington, DC: Government Printing Office, 1893), 452.

Chapter 16: Racing Time

147   The schooner Hans helped: Edward H. M. Davis, *Papers Respecting the Declaration of a British Protectorate Over the Gilbert Islands*, Foreign Office Confidential Print 6269 (1892), 146.

148   Like John on *Ebon*: Davis, *Papers*, 145; "The Blackbird Cruise," *San Francisco Examiner*, October 16, 1892; "A Sale of Souls," *San Francisco Examiner*, October 15, 1892.

148   Rumors circulated: "Sons of Captain Walker Refute Wandering Minstrel Story," *Honolulu Advertiser*, January 30, 1922; Davis, *Papers*, 141–44.

149   Ferguson was notorious: Charles H. Shinn, "Only South-Sea Islanders," *The Friend* (Honolulu), March 20, 1897; "Here's a Pretty Scheme," *Hawaiian*

Star, July 25, 1893; David McCreery and Doug Munro, "The Cargo of
the Montserrat: Gilbertese Labor in Guatemalan Coffee, 1890–1908," *The
Americas* 49, no. 3 (1993): 273; "A Sale of Souls," *San Francisco Examiner*,
October 15, 1892.

150 Now that he was approached: Arthur Inkersley, "A Trip to the Kingswell Islands,"
Travel: An Illustrated Monthly Magazine, January 1896, 443.

150 Hans agreed: "A Sale of Souls," *San Francisco Examiner*, October 15, 1892.

151 Returning to *Montserrat*: "The Blackbird Cruise," *San Francisco Examiner*,
October 16, 1892; "A Sale of Souls," *San Francisco Examiner*, October 15, 1892;
Davis, *Papers*, 142.

151 Hans's circumstances: Edward Hertslet, ed., *A Collection of Treaties and
Conventions, Between Great Britain and Foreign Powers, and of the Laws,
Decrees, Orders in Council*, vol. 19 (London: Butterworth & Co., 1895), 569;
"Experiences of a 'Blackbirder' among the Gilbert Islanders," *Overland Monthly*,
June 1894, 571–72.

152 Throughout a number: Admiralty Officers Service Records, Edward Henry
Meggs Davis, National Archives.

152 He was fascinated: Admiral Edward Henry Meggs Davis Collection, British
Museum, items Oc1903,-.145 (bowl), Oc1894,-.254 (whale tooth ornament),
Oc1904,0621.13 (coffin); Alison Clark, *Resonant Histories: Pacific Artefacts and
the Voyages of the HMS Royalist 1890–1893* (Leiden, Netherlands: Sidestone
Press, 2019), 81, 86.

152 Davis embraced his role: Davis, *Papers*, 248; Clark, *Resonant*, 41.

152 Captain Davis heard: Robert Louis Stevenson to Sidney Colvin, June 1889,
in *The Letters of Robert Louis Stevenson*, ed. Sidney Colvin, vol. 3 (London:
Methuen & Co., 1911), 130; Davis, *Papers*, 21; "Local & General News," *Daily
Bulletin* (Honolulu), December 9, 1887.

153 Beginning an investigation: Davis, *Papers*, 145–46.

153 The captain called: "The Blackbird Cruise," *San Francisco Examiner*, October
16, 1892; Davis, *Papers*, 146.

154 Though Hans's blackbirding: "120 Died on the Black Doctor's Slaughter Ship,"

Truth (Sydney), January 22, 1950; "The Blackbird Cruise," *San Francisco Examiner*, October 16, 1892.

154   Hans could not: Davis, *Papers*, 138.

155   Hans was tried: "The Blackbird Cruise," San Francisco Examiner, October 16, 1892; Clark, *Resonant*, 263, n. 38.

155   During Hans's two-day trial: Davis, *Papers*, 138–44.

156   Hans was blindfolded: "Sons of Captain Walker Refute Wandering Minstrel Story," *Honolulu Advertiser*, January 30, 1922.

157   J. L.'s doggedness: "Pickpockets in a Foreign Auction Room," *Japan Weekly Mail*, November 25, 1893; "Shooting Affray in Yokohama," *Japan Weekly Mail*, July 29, 1893 (another auction incident).

157   Family deaths: "Destructive Fire on the Bluff," *Japan Weekly Mail*, August 12, 1893.

157   J. L.'s firm: "Total Wreck of the Barque 'Alette,'" *Japan Weekly Mail*, September 18, 1897; Mary Chipman Lawrence, *The Captain's Best Mate: The Journal of Mary Chipman Lawrence on the Whaler Addison, 1856–1860*, ed. Stanton Garner (Providence, RI: Brown University Press, 1966), 16.

158   By this time: "Death," *Japan Weekly Mail*, October 4, 1890.

158   Japan had a rich tradition: Raphael Pumpelly, *Across America and Asia: Notes of a Five Years' Journey Around the World, and of Residence in Arizona, Japan, and China* (New York: Leypoldt & Holt, 1870), 83–84; Madeline Anne Wallace-Dunlop, *Glass in the Old World* (London: Field & Tuer, 1822), 221–22; "Fortune Telling in Japan," *Frank Leslie's Popular Monthly*, October 1878, 440.

158   J. L. viewed himself: "Death of Mr Barber Field," *Japan Weekly Mail*, July 28, 1888.

159   The fortune-tellers: "Fortune Telling in Japan," *Frank Leslie's Popular Monthly*, October 1878, 440; Nicola Liscutin, "Mapping the Sacred Body: Shinto Versus Popular Beliefs at Mt. Iwaki in Tsugaru," in *Shinto in History: Ways of the Kami*, ed. John Breen and Mark Teeuwen (New York: Routledge, 2009), 190; Takefusa Sasamori, "Therapeutic Rituals Performed by Itako," *World of Music* 39, no. 1 (1997): 85–96; Ichirō Hori, "Shamanism in Japan," *Japanese Journal of Religious Studies* 2, no. 4 (December 1975), 231–87.

159   At times: Jason Ānanda Josephson, *Invention of Religion in Japan* (Chicago: University of Chicago Press, 2012), 181.

159   The soothsayer Kiyo: Elizabeth Walker, "Wrecked on Midway Island in 1888," *Paradise of the Pacific*, November 1936, 28.

159   Kiyo begged J. L.: "Norma Has a History," *Hawaiian Star*, August 24, 1900; A. Farsari, *Keeling's Guide to Japan* (Yokohama: A. Fasari, 1890), 47; "At the Bay," *Sacramento Union*, January 26, 1889.

160   Another possibility: "Tale of a Smuggler," *Seattle Post Intelligencer*, December 11, 1892; "Sold Under the Hammer," *Hawaiian Star*, September 25, 1899; "There Was Launched," *Japan Weekly Mail*, October 13, 1888.

161   Charles Johnson, however: "A Bold Yankee Pirate," *Weekly Courier* (Jasper, Indiana), April 20, 1888.

161   Charles, like many other: "Tale of a Smuggler," *Seattle Post Intelligencer*, December 11, 1892; Andrew Farrell, "Island Wrecks VII," *Pacific Marine Review*, November 1920, 58; "How Schooner Norma Landed Her Opium Cargo," *Honolulu Advertiser*, February 13, 1901.

161   When J. L. had the chance: Walker, "Wrecked," 28; "Survivors Now Here," *Honolulu Advertiser*, August 15, 1900.

Chapter 17: Letter in a Bottle

163   The Walker family knew: "The Wandering Minstrel," *San Francisco Chronicle*, July 1, 1889; Elizabeth Walker, "Wrecked on Midway Island in 1888," *Paradise of the Pacific*, November 1936, 28; Mary Chipman Lawrence, *The Captain's Best Mate: The Journal of Mary Chipman Lawrence on the Whaler Addison, 1856–1860*, ed. Stanton Garner (Providence, RI: Brown University Press, 1966), 3.

163   Elizabeth and the boys: "The Wandering Minstrel," *San Francisco Chronicle*, July 1, 1889; "Norma Has a History," *Hawaiian Star*, August 24, 1900.

164   On March 17: Frederick D. Walker, *Log of the Kaalokai* (Honolulu: Hawaiian Gazette Co., 1909), 57.

164   As he chopped: Walker, *Log*, 57; George Read, *The Last Cruise of the Saginaw* (Boston: Houghton Mifflin, 1912), 86–87.

164   Experiencing the same mixture: "The Mystery Solved," *Daily Pacific Commercial Advertiser*, June 13, 1889.

164   As they had learned: "The Wandering Minstrel," *San Francisco Chronicle*, July 1, 1889.

165   The sail they had spotted: Hyman Kublin, "The Discovery of the Bonin Islands: A Reexamination," *Annals of the Association of American Geographers* 43, no. 1 (1953): 32, 43–46; James Cumming Dewar, *Voyage of the Nyanza* (Edinburgh: William Blackwood and Sons, 1892), 327.

166   That story of the treasure: Edgar K. Thompson, "The Lost Treasure of Agrigan," *United States Naval Institute Proceedings* 71, (February 1945): 209–12.

166   Decades later: Events in Charles Johnson's treasure hunt drawn for this section from "Piracy and Hidden Treasure," *Japan Weekly Mail*, April 28, 1888; "A Bold Yankee Pirate," *Weekly Courier* (Jasper, Indiana), April 20, 1888; "Information," *Japan Weekly Mail*, December 1, 1888.

168   Now a revitalized Charles was back: "Wandering Minstrel," *Honolulu Advertiser*, April 8, 1889; "Tale of a Smuggler," *Seattle Post Intelligencer*, December 11, 1892.

168   They remained anchored: "Wandering Minstrel," *Honolulu Advertiser*, April 8, 1889; "Tale of a Smuggler," *Seattle Post Intelligencer*, December 11, 1892; "Serious Assault," *Honolulu Advertiser*, January 7, 1889.

169   He also had the voice: "Norma Has a History," *Hawaiian Star*, August 24, 1900.

169   Now that they had seen: "The Wandering Minstrel," *San Francisco Chronicle*, July 1, 1889; Read, *Last Cruise*, 88; Walker, "Wrecked," 28; Walker, *Log*, 22.

170   After fourteen months: "Cast on a Desert Island," *Custer Chronicle* (Custer, South Dakota), May 25, 1889; "The Wandering Minstrel," *San Francisco Chronicle*, July 1, 1889.

170   The American schooner captain: "Norma Has a History," *Hawaiian Star*, August 24, 1900.

171   Robert Louis Stevenson and Lloyd Osbourne: Robert Louis Stevenson and Lloyd Osbourne, *The Wrecker* (London: Cassell, 1892), 386.

171   The character of Trent: Stevenson and Osbourne, *Wrecker*, 389–90.

171   Similarly, Charles claimed: "Norma Has a History," *Hawaiian Star*, August 24, 1900; "Tale of a Smuggler," *Seattle Post Intelligencer*, December 11, 1892.

172   Larger opportunity emerged: Dewar, *Voyage*, 393–94; "Miscellaneous," *Honolulu Advertiser*, January 24, 1889; "Steamers Arrived," *Boston Evening Transcript*, February 9, 1889; "Trouble in San Francisco," *Buffalo Times*, January 9, 1889; "Wan Tong Quarantined," *Vancouver Daily Times*, January 8, 1889; "China and Corea," *Independent-Record* (Helena, Montana), January 8, 1889.

173   Stormy weather: "Wandering Minstrel," *Honolulu Advertiser*, April 8, 1889.

174   As Elizabeth had managed: "Norma Has a History," *Hawaiian Star*, August 24, 1900; "Wandering Minstrel," *Honolulu Advertiser*, April 8, 1889.

174   At the hut: Lyle S. Shelmidine, "The Early History of Midway Islands," *American Neptune*, July 1948, 187.

175   In the morning: Walker, *Log*, 57.

Chapter 18: *Wreckages and Renewals*

176   As the *Norma* set sail: "Wandering Minstrel," *Honolulu Advertiser*, April 8, 1889; "On a Desert Island," *San Francisco Chronicle*, April 21, 1889; Abby Jane Morrell, *Narrative of a Voyage to the Ethiopic and South Atlantic Ocean, Indian Ocean, Chinese Sea, North and South Pacific Ocean, in the years 1829, 1830, 1831* (New York: Harper, 1833), 20.

176   With clear weather: "Local & General News," *Evening Bulletin* (Honolulu), April 8, 1889; "Wandering Minstrel," *Honolulu Advertiser*, April 8, 1889; "On a Desert Island," *San Francisco Chronicle*, April 21, 1889.

178   Elsewhere, John Cameron's shock: "Disasters at Sea," *Evening Chronicle* (Newcastle upon Tyne, England), May 2, 1889.

179   In a last twist: Wreck Report, "Wandering Minstrel," Port Cities Southampton, No. 3792; Homer C. Votaw, "Midway—The North Pacific's Tiny Pet," *United States Naval Institute Proceedings* 66, no. 453 (November 1940): 43.

179   While the Walkers: "Local & General News," *Evening Bulletin* (Honolulu), August 17, 1889; "Death of George Hanker," *Hilo Daily Tribune*, December 13, 1901.

180   Reporters in Honolulu: R. C. Terry, introduction to *Robert Louis Stevenson: Interviews and Recollections* (London: Macmillan, 1996), xvii; Harry Jay Moors, *With Stevenson in Samoa* (London: Adelphi Terrace, 1910), 3.

180   Stevenson thought: Robert Louis Stevenson to Sidney Colvin, June 1889, in *The Letters of Robert Louis Stevenson*, ed. Sidney Colvin, vol. 3 (London: Methuen & Co., 1911), 153–54; Graham Balfour, *The Life of Robert Louis Stevenson* (London: Methuen & Co., 1911), 271.

181   Stevenson recruited: "The Midway Island Cable Station," *San Francisco Chronicle*, August 2, 1903; Frederick Chamberlin, *Around the World in Ninety Days* (Boston: Clark Publishing, 1906), 87; "Survivors Now Here," *Honolulu Advertiser*, August 15, 1900; Fanny Stevenson, preface to *The Wrecker*, Biographical Edition, by Robert Louis Stevenson and Lloyd Osbourne (1905: Scribner's, 1912), vii.

181   When the Stevensons: Fanny Stevenson, *The Cruise of the "Janet Nichol" Among the South Sea Islands* (New York: Scribner's, 1914), 120; John Cameron, "First Mate Cameron's Story of the Wreck, His Sojourn on Midway and His Voyage to the Marshalls," *Pacific Marine Review*, January 1922, 37.

182   Everyone involved: "Sons of Captain Walker Refute Wandering Minstrel Story," *Honolulu Advertiser*, January 30, 1922.

182   In addition to the nine: Homer C. Votaw, "Midway—The North Pacific's Tiny Pet," *United States Naval Institute Proceedings* 66, no. 453 (November 1940): 43.

182   These references: Wreck Report, "Wandering Minstrel," Port Cities Southampton, No. 3792.

183   To the extent: Robert Louis Stevenson and Lloyd Osbourne, *The Wrecker* (London: Cassell, 1892), 392.

184   The Walkers recuperated: Frederick D. Walker, *Log of the Kaalokai* (Honolulu: Hawaiian Gazette Co., 1909), 4; "Charles D. Walker Dies; Collector, Boat-Builder," *Honolulu Star Bulletin*, May 2, 1960; Lyle S. Shelmidine, "The Early History of Midway Islands," *American Neptune*, July 1948, 187.

184   Meanwhile, the Walker boys: "Sons of Captain Walker Refute Wandering Minstrel Story," *Honolulu Advertiser*, January 30, 1922.

184   Rather than wait: "Storm-Scourged Ships," *San Francisco Examiner*, December 16, 1894; "Sons of Captain Walker Refute Wandering Minstrel Story," *Honolulu Advertiser*, January 30, 1922; "Survivors Now Here," *Honolulu Advertiser*, August 15, 1900.

185   At one point: John Cameron, transcribed by Andrew Farrell, *John Cameron's Odyssey* (New York: Macmillan Company, 1928), 341–49, 390; "Flew a Black Flag," *Evening Bulletin* (Honolulu), April 12, 1894; "From South Seas," *Hawaiian Gazette*, December 4, 1896; Walker, *Log*, 418.

185   At strategic points: Cameron, *Odyssey*, 271; "Sons of Captain Walker Refute Wandering Minstrel Story," *Honolulu Advertiser*, January 30, 1922; Andrew Farrell, "Island Wrecks VII," *Pacific Marine Review*, November 1920, 57.

186   Mouwen, who had fled: Fanny Stevenson, *The Cruise of the "Janet Nichol" Among the South Sea Islands* (New York: Scribner's, 1914), 120; "Loss of the Schooner Tiernan," *San Luis Obispo Tribune* (San Luis Obispo, CA), March 8, 1890; "The U.S.S. Iroquois," *Hawaiian Gazette*, April 1, 1890.

186   The array of insurance agents: *Insurance Times*, June 1890, 289; "The South British Fire and Marine Company," *Taranki Herald* (Taranki, New Zealand), June 24, 1890.

187   Long after the unsuccessful search: Flint's Pharmacy, advertisement, *Oakland Tribune*, July 13, 1897.

187   USS *Alliance*'s Henry Clay Taylor: "Henry Clay Taylor," in *Dictionary of American Naval Fighting Ships*, ed. James L. Mooney, vol. 7 (Washington, DC: Naval Historical Center, 1981), 69; "Admiral's Fatal Fall," *Taunton Courier and Western Advertiser*, October 16, 1929.

187   The steamship captain: "Survivors Now Here," *Honolulu Advertiser*, August 15, 1900; "The Mongolia Is in Bad Position," *Hawaiian Star*, September 17, 1906; "Died from Accident," *Fall River Daily Evening News*, December 27, 1906.

187   Captain Henry Stap: "Sarah Agnes Stap," in *Encyclopedia Titanic*, accessed April 20, 2024, https://www.encyclopedia-titanica.org/titanic-survivor/sarah-agnes-stap.html; "Halewood SV (1885–1913) Songvig SV (+1920)," in *Wreck Site*, accessed April 20, 2024, https://wrecksite.eu/wreck.aspx?237383.

188   Another ship: "Hurricane at Samoa," *Auckland Star*, March 30, 1889.

188   The vessel that closed in: "The Shaw, Savill and Albion Company," *Auckland Star*, December 22, 1890; "Last evening," *Auckland Star*, October 27, 1891; "Convict Casey," *San Francisco Chronicle*, March 30, 1893; "Police Court—This Day," *Auckland Star*, January 16, 1891; "Double Shotted Guns," *Bangor Daily Whig and Courier*, September 11, 1894; "Those 'Blarsted' Yankees," *Portland Daily Press* (Maine), September 12, 1894.

189   The only captain: "Report of Wreck," National Archives of the United Kingdom, BT 15/30/F4673; "Tale of a Smuggler," *Seattle Post Intelligencer*, December 11, 1892; "Wharf and Wave," *Honolulu Advertiser*, April 11, 1893.

189   Charles's former first mate: "The story related," *Japan Weekly Mail*, May 5, 1888; "50 Great Treasure Islands," *Islands Magazine*, July/August 1994, 128; George Hemenes, "Treasure Chests," *Paradise of the Pacific*, 1959, 113.

189   Frederick Walker still dreamed: "To Search for Buried Treasure," *Evening Bee* (Honolulu), August 4, 1903; "A Honolulu Man Knows Where Much Silver Has Been Buried," *Honolulu Advertiser*, July 6, 1903; "Charles D. Walker Dies; Collector, Boat-Builder," *Honolulu Star Bulletin*, May 2, 1960.

190   A legend spread: "Charles D. Walker Recalls Building Thousands of Boats," *Honolulu Star Bulletin*, December 9, 1953; "Survivors Now Here," *Honolulu Advertiser*, August 15, 1900; "A Honolulu Man Knows Where Much Silver Has Been Buried," *Honolulu Advertiser*, July 6, 1903; Walker, *Log*, 16, 55; Shelmidine, "Early History," 188; Walter Rothschild, *The Avifauna of Laysan and the Neighboring Islands with a Complete History to Date of the Birds of the Hawaiian Possessions* (London: R. H. Porter, 1893–1900), vii.

190   Even before: Walker, *Log*, 54.

191   "We felt a sickening feeling: Walker, *Log*, 54, 61; Shelmidine, "Early History," 186.

191   One of the naturalists: Shelmidine, "Early History," 188.

191   Freddy, meanwhile: "Pacific Views," *Daily Commercial Pacific Advertiser*, November 23, 1891.

191   The naturalists complained: Walker, *Log*, 54, 57.

192   They had traveled: Alexander MacDonald, "Down the Gang Plank," *Honolulu*

Star Bulletin, March 6, 1934; "Old Cannon, Taken from Pirate Junk, Is Recovered After Wreck at Midway," *Honolulu Advertiser*, April 22, 1923.

192   The Walkers soon prospered: "Wharf and Wave," *Honolulu Advertiser*, April 11, 1893; "Marine Notes," *Hawaiian Star*, May 2, 1893.

192   Elizabeth never showed: Elizabeth Walker, "Wrecked on Midway Island in 1888," *Paradise of the Pacific*, November 1936, 27–28.

193   Robert Louis Stevenson's determination: Matthew Pearl, "Where Did Kurtz Come From?," *Slate*, April 29, 2015, https://slate.com/culture/2015/04/kurtz -and-heart-of-darkness-did-joseph-conrad-base-mr-kurtz-on-robert-louis -stevenson.html.

193   Upon first hearing: Robert Louis Stevenson to Sidney Colvin, June 1889, in *The Letters of Robert Louis Stevenson*, ed. Sidney Colvin, vol. 3 (London: Methuen & Co., 1911), 153–54; "Survivors Now Here," *Honolulu Advertiser*, August 15, 1900; Robert Irwin Hillier, *The South Seas Fiction of Robert Louis Stevenson* (New York: Peter Lang, 1989), 83.

194   Secretary of State Thomas Bayard was lampooned: "Notes of the Day," *Burlington Nonpareil* (Kansas), March 1, 1889; Nancy J. Morris, introduction to *The Biography of Hon. Robert William Wilcox*, by Thomas K. Nananaela (Honolulu: privately published, 1993), xi; Thomas K. Nananaela, *The Biography of Hon. Robert William Wilcox*, trans. Nancy J. Morris (Honolulu: privately published, 1993), 20–26; "James Hay Wodehouse Made History Once," *Maui News*, October 18, 1913; Liliuokalani, *Hawaii's Story by Hawaii's Queen* (Boston: C. J. Peters & Son, 1898), 201; David Star, "The Honorable Robert Kalanihiapo Wilcox," Royal Family of Hawaii Official Website, accessed April 21, 2024, https://www.crownofhawaii.com/robert-kalanihiapo-wilcox.

194   American political, business, and commercial: William Reynolds to H. K. Thatcher, September 30, 1867, in *Senate Documents, Otherwise Publ. as Public Documents and Executive Documents 14th Congress, 1st Session-48th Congress, 2nd Session and Special Session* 7 (Washington, DC: Government Printing Office, 1868), 7.

195   Frederick complimented: Walker, *Log*, 63; "Aided by British," *San Francisco Examiner*, January 20, 1895; "On a Mid-Ocean Island," San Francisco Call,

February 26, 1895; *Papers Relating to the Foreign Relations of the United States* (Washington, DC: Government Printing Press, 1871), 587.

196 Edvart Olsen, one of the survivors: "Olsen Service Slated Today," *Honolulu Advertiser*, January 10, 1939; Walker, *Log*, 53, 63.

196 The Walkers remained: Carl Elschner, *The Leeward Islands of the Hawaiian Group: Contributions to the Knowledge of the Islands of Oceania* (Honolulu: *Sunday Advertiser*, 1915), 61; Gerrit P. Wilder, "A Short Trip to the Midway Islands with Captain A. P. Niblack in the U.S.S. 'Iroquois,'" *Hawaiian Forester and Agriculturist* 2, no. 12 (December, 1905): 394; George W. Cox, *Alien Species in North America and Hawaii: Impacts on Natural Ecosystems* (Washington, DC: Island Press, 1999), 26; "Iroquois Is Here Again," *Hawaiian Star*, August 6, 1900; "Cable Station at Midway," *Hawaiian Star*, August 6, 1900; "The American Cable Across the Pacific," *Telegraph Age*, July 16, 1903, 360; "Tiny Islets Nerve Center for Messages," *Honolulu Advertiser*, July 27, 1932.

197 Elizabeth and Frederick died: "F. G. E. Walker Died; Survived Midway Shipwreck," *Honolulu Star Bulletin*, December 1, 1953; "Charles D. Walker Dies; Collector, Boat-Builder," *Honolulu Star Bulletin*, May 2, 1960; MacDonald, "Down the Gang Plank."

198 Shipwrecks make an impact: K. E. D. Coan, "New Study Finds 19th Century Wooden Shipwrecks to Be Thriving Habitats for Deep-Sea Microbiomes," *Frontiers Science Communications*, June 8, 2022, https://blog.frontiersin .org/2022/06/08/new-study-finds-19th-century-wooden-shipwrecks-to-be -thriving-habitats-for-deep-sea-microbiomes/; "The British Sailing Vessel *Dunnottar Castle*," Papahānaumokuākea Marine National Monument, accessed April 21, 2024, https://www.papahanaumokuakea.gov/maritime/dunnottar .html; for examples of surveys of submerged sites around Midway, see "2003 Kure and Midway Atoll Heritage Survey," National Marine Sanctuaries, accessed April 22, 2024, https://sanctuaries.noaa.gov/maritime/expeditions /nwhi_2003.html, and Kelly Keogh, "Mission Plan," NOAA Ocean Exploration, accessed April 22, 2024, https://oceanexplorer.noaa.gov /explorations/17midway/background/plan/plan.html.

198   Within a decade: Ralph Linwood Snow and Douglas K. Lee, *A Shipyard in Maine: Percy & Small and the Great Schooners* (Maine: Tilbury House, 199), 287; "Disasters at Sea," *Evening Chronicle* (Newcastle upon Tyne, England), May 2, 1889; "Awfully Utter," *Cincinnati Enquirer*, January 3, 1882.

199   As for sharking: "Shark Oil Good, Minus the Smell," *Honolulu Advertiser*, April 20, 1912.

200   The number of birds: Connie Toops, "Midway—It's for the Birds," *The Yellowhammer*, Fall 2012, 10; Thomas Heaton, "Plan to Eradicate Mice That Attack Albatrosses on Midway Delayed Another Year," *Honolulu Civil Beat*, March 28, 2022, https://www.civilbeat.org/2022/03/plan-to-eradicate-mice-that-attack-albatross-on-midway-delayed-another-year/; Aaron Ochoa, "How Removing Rats Helped to Save Midway's Seabirds," *Island Conservation*, February 14, 2018, https://www.islandconservation.org/removing-rats-saved-midways-seabirds/; George C. Munro, "The Small Birds of Midway," *Elepaio* 6 (1945): 13–14.

201   Midway is littered: Nick Paton Walsh, Ingrid Formanek, Jackson Loo, and Mark Phillips, "Plastic Island," CNN.com, accessed April 21, 2024, https://www.cnn.com/interactive/2016/12/world/midway-plastic-island/.

201   Honolulu-born president: "Global Sea Level Could Rise 50 Feet by 2300, Study Says," Rutgers University, accessed April 21, 2024, https://www.rutgers.edu/news/global-sea-level-could-rise-50-feet-2300-study-says; Maria Gallucci, "Obama Visits Remote Midway Atoll to Highlight Climate Change Threats," Mashable, September 2, 2016, https://mashable.com/article/obama-midway-atoll-climatechange; Nick Thompson, "Midway: Why Barack Obama Visited a Tiny Island in the Pacific," CNN.com, September 2, 2016, https://www.cnn.com/2016/09/01/politics/midway-obama-preview/index.html.

202   In a preface: Fanny Stevenson, preface, *The Wrecker*, Biographical Edition, by Robert Louis Stevenson and Lloyd Osbourne (New York: Scribner's, 1905), viii; "From South Seas," *Hawaiian Gazette*, December 4, 1896.

202   Over the years: Walker, "Wrecked," 27–28.

INDEX

ABOUT THE AUTHOR

MATTHEW PEARL's books have been international and *New York Times* bestsellers, and have been translated into more than thirty languages. His nonfiction writing has appeared in the *New York Times*, the *Boston Globe*, and *Slate*, and he edits *Truly*Adventurous* magazine. He has been chosen as Best Author in *Boston* magazine's "Best of Boston" issue and received the Massachusetts Book Award for Fiction.